-4. SEP. 1986

LLYFRGELL
YNYSYBWL
LIBRARY

AF841112

CYNON VALLEY BOROUGH LIBRARIES
BWRDEISTREF CWM CYNON

LLYFRGELLOEDD BWRDEISTREF SIROL
RHONDDA-CYNON-TAF
DISCARDED
DIFETHWYD
RHONDDA-CYNON-TAFF
COUNTY BOROUGH LIBRARIES

4/52

THE FIRST STONEHENGE

THE FIRST STONEHENGE

Gaynor Francis

CYNON VALLEY BOROUGH
LIBRARIES
LLYFR... ...EDD BWRDEISTREF
... CYNON

PLEASE ...
Library b... ... kept beyond
the time a... ...sion of time
may be gr... ...for extension
need not ... ...the books,
provi... ...names
of the authors, titles an... ...ers of the
books, and dates due ... the case
of non-fiction ... also necessary to give
class numbers.
Overdue ... ...ther fines;
These fines ... ...orceable by law.
Please keep boo... ...ean and protected
from damage ... Rain ruins books
In case of contact with infectious disease,
please inform the ...vironmental Health
Offi...

Christopher Davies

Copyright © Gaynor Francis 1986

Published by
Christopher Davies (Publishers) Ltd.,
P.O. Box 403, Sketty,
Swansea, SA2 9BE.

All rights reserved. No part of this publication may be reproduced, stored in a retrieval system, or transmitted, in any form or by any means, electronic, mechanical, photocopying, recording or otherwise, without the prior permission of Christopher Davies (Publishers) Ltd.

930.13
F8

ISBN 0 7154 0666 3

*Typeset in 13/15 point Bembo
and printed in Wales by
Dynevor Printing Company
Rawlings Road
Llandybïe, Dyfed*

253922

To
The University College of Swansea
this book is dedicated

alma mater mea

Acknowledgements

I have first to acknowledge the influence of my parents, who introduced me to mountain paths at a very early age although life was to take me in quite another direction, and the influence of Lionel, who was to lead me back. To Emeritus Professor T. R. Owen whose infectious enthusiasm, practical advice and discreet lack of pressurisation helped me bring this book to completion; to Dr Derek Maling for reassurance mathematical; to Gerald Thomas Esq. (ex-Royal Navy and Merchant Navy) for imparting knowledge and skills therefrom acquired. To the farmers and landowners who so readily gave me permission to cross their land and who, so often, gave their hospitality as well; and in particular to the Mansel Davies family of Llanfyrnach whose enthusiastic help and local knowledge smoothed so considerably the Preseli Standing Stone study. To Mr G. B. Lewis for his most able assistance and patience in reproducing very complicated diagrams, to Mr D. E. Price for photographic help, to Miss Carole Morse for text reading and to Mrs D. J. Nuttall for typing the manuscript.

Gaynor Francis

Contents

List of Illustrations

Figure No. *Page*

List of Plates

Aros mae'r mynyddau mawr,
Rhuo trostynt mae y gwynt;
Clywir eto gyda'r wawr
Gân bugeiliaid megis cynt.
Eto tyfa'r llygad dydd
O gylch traed y graig a'r bryn,
Ond bugeiliaid newydd sydd
Ar yr hen fynyddoedd hyn.

Ceiriog

Foreword

It is with much pleasure that I commend this book to the reader. Archaeology and the activities of Early Man have always been a matter of interest and curiosity to me. Two questions have, however, never failed to trouble me.

Firstly, what about Man in Britain, and especially in Wales, *before* the New Stone Age? Much is known about Neolithic times and later, but far less has been published on Pre-Neolithic (particularly Mesolithic) Man and his activities in Wales.

Secondly, while most people are agreed that the famous "bluestones" of Stonehenge probably came from the Preseli Hills in Pembrokeshire, no one has really answered my question, "Why did they take them from that far corner of Wales? Dartmoor, the Malverns and the Mendips — those regions were much nearer."

I am of the opinion that in this book, the writer has gone a long way towards answering these questions.

I have accompanied her on several of her field study expeditions and I was each time struck by her familiarity with and close knowledge of even the most isolated and remote of tracts in the not inconsiderably large area that the Welsh Uplands cover.

This knowledge, combined with the careful and detailed way in which she has developed her thesis, has enabled her to succeed in producing a book that can kindle the interest both of the man in the street and the academic, a book that leaves the reader more thoughtful about the world about him, if he is a walker on the hills — and if he is not, then it leaves him with something of significance to think about.

T. R. Owen
(Emeritus Professor of Geology)

Preface

This book was originally intended to be purely an account of walks that I had undertaken in the hill-country of Central Wales drawing attention to certain repetitive features that I had noticed and that had fascinated me.

The Mesolithic peoples who lived in that area at the time in question played an integral part in the story and from the moment that they walked in 'live on stage', as it were, they took over the tale and made the extension of the original parameters unavoidable, if full justice were to be done to themselves and to the part that they played in this heritage of ours. Chapters I, II, III came into being for this reason, and for Pam, who suffered a dilemma, together with many other of my friends who suffered similarly, I have written these particular chapters.

I hope that from Chapter IV on, the reader will be able to identify with those early Mesolithic peoples as they set themselves the task of appointing hill-tops which would qualify as acceptable alignments to a point on nature's far distant horizon. A point on the horizon from behind which the sun could be expected to appear at a certain time of the year, thus giving them an indication of where they stood on the pathway of the year's seasons. The degree of accuracy astounds when bearing in mind the primitive stratagems resorted to, and while inaccuracies may be evident to the trained eye of Modern Man, then he must remind himself that Ancient Man's aligning hill-top system worked well enough in the light of his own particular needs.

The fact that the Sun does not now rise exactly from the same point on the horizon as it did in Mesolithic times at a given time, would not seem, in view of the circumstances given here, a matter to be taken too much into account as it still does so approximately enough. Nor, in my view, is the question of the values of the azimuth of the midsummer sunrise relative to the differing degree of latitude between Stonehenge and, let us say, a point on the Brecon Beacons worthy of too much concern. As Dr D. H. Maling (in personal communication) points out, "these figures", (meaning those resulting from a comparison of the relative latitudes and their modern azimuth and their 2,000 B.C. azimuth measurements), "show that there is only a difference of a few minutes in arc between the azimuths of the rising sun at the two places and this is about the same order of magnitude in the difference between the values computed for 2,000 B.C. and the present". Dr Derek Maling says that, for the comparison of the azimuths, the method he used was the Haversine Formula of spherical Trigonometry and for such expert clarification on this point, I am much indebted to Dr Maling.

As, then, the primitive methods of Ancient Man produced a result that was satisfactory to him, then Modern Man can do nothing other than accept it in that spirit, while, at the same time, according the highest respect and honour to this unexpectedly able forebear who laid the foundation of what is today the very complicated and highly sophisticated mathematical science of astronomy.

Gaynor Francis
Gower, West Glamorgan, 1.7.85

CHAPTER I

Who were they and when did they live?

This is about people. It is more about people than about places. Very real, but hard-to-imagine people. Hard to imagine because they lived between about 8,000 and 4,000 years before Christ. It is these people who lived as the last Devensian Ice Age was ending and the promise of a tolerable life was beginning. And it is these people who gave the places where they lived the importance that they had.

The time-scale of 8,000 B.C. to 4,000 B.C. already begins to stretch the ability of the human mind to conceive, being almost four times as long before the birth of Christ as man has lived after it.

Largely for the reason that their primitive form of existence, in the area in which we intend to walk, would seem to have left no substantial trace of their passing, coupled with the great passage of time, far less enquiry has been made into the way of living of these Mesolithic peoples than has been made into that of the Neolithic peoples who followed them. As a direct result of the changed habits of living, much more has been left behind by the Neolithic peoples thus making an assessment of their life-style so much easier.

The Mesolithic period is what the length of time between the immediate post-glacial period of about 9,000 B.C. up until about 4,000 B.C. is called.

These Mesolithic peoples were the descendants of those who had lived in the Late Upper Palaeolithic period — the late glacial period of approximately 13,000 to 10,000 B.C. This produced a climate of subarctic conditions as was the case in the areas of the part of Britain which interest us.

It is what is known as the Devensian Ice Age (or the Last Ice Age)

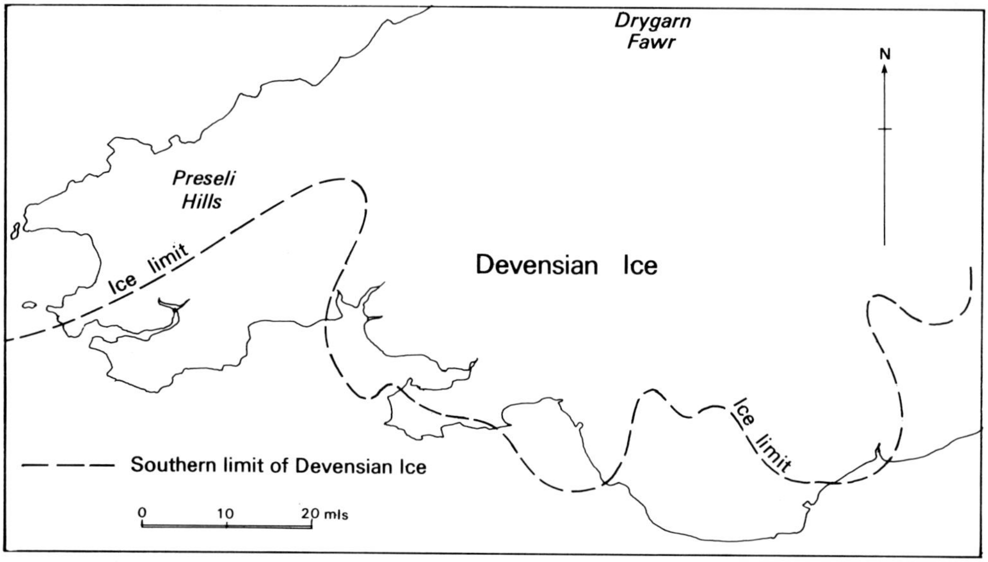

Fig. 1. *Southern limit of the last (Devensian) glaciation in South Wales (after Bowen). The coastline is shown as it is today but in those days, owing to lower sea-levels, it was possible to cross the present-time Bristol Channel and Carmarthen Bay on foot.*

that was responsible for this glacial climate, the approximate most southerly advanced line of which (see Fig. 1) still left much of what is known today as South Pembrokeshire and South Gower free of sheet ice but still, nevertheless, an area in which tundra conditions prevailed.

To be remembered too, is that sea-levels in this late glacial period were very low, as the later rising waters of the post-glacial period were still imprisoned only a short distance farther to the north and to the east in the form of ice, hundreds of feet thick. The resulting effect of this low sea-level meant that what are today the sea areas of Carmarthen Bay, the Bristol Channel and the Straits of Dover were, in the glacial period, dry land — thus affording virtually unimpeded access, as far as glacial ice was concerned, to man and animal from as far away as France across Cornwall, Devon and all the way on to Pembrokeshire.

In large areas of France, where Early Man suffered no similar set-back from the worst effects of the Devensian Ice Age, he was enabled to develop and he produced, for this reason, more advanced forms of art — such as cave drawings of remarkable beauty, the like of which have not so far been found in the glacial ice fringe areas of southernmost Britain where Man's uncomfortable life and consequent struggle to survive usurped all other interests. In his caves, this Man left piles of bones.

Since the majority of finds in the Britain of this late glacial period are from caves (as, for example, those from caves in the Pembroke, Tenby and Caldey areas — see Chapter XIV) it would tend to indicate that caves as opposed to open sites were virtually the only shelter in the majority of areas against the harshness of the prevailing climatic conditions.

The vegetation of the tundra of the time, while not capable of being exploited by Man for any leaf and berry nourishment to any great extent, was, however, ideally suited to the needs of the reindeer and judging from the evidence of finds of enormous piles of bones, antlers and, in some cases, implements made from them, it would seem that reindeer was the staple diet of Man in those late glacial times and was for this reason of the utmost importance in the life of the Upper Palaeolithic peoples.

J. G. Evans in "*The Environment of Early Man in the British Isles*" (1975) says that evidence from Tornewton cave in Devon shows a seasonal occupation by reindeer hunters in the Spring and early Summer and that "there are sites for example in the Dordogne where there is a clear year-round pattern of reindeer exploitation."

Thus, with evidence of the seasonal exploitation of herds already a well-established hunting custom of Upper Palaeolithic times, it would seem only natural that such a customary method, as was this seasonal exploitation of the herds, should be found to have been carried through into the way of life of their descendants of the post-glacial period. No longer were the herds the reindeer of the sub-arctic tundra but the deer of the forests of the warmer post-glacial period. Moreover, as the need of pre-cognisance and recognition of the seasons still remained, so would it seem only natural that the methods long used to attain this knowledge would similarly have been carried forward and the long revered appearances of the early February, early May, early August and early November sunrises and sunsets, so important to the people of the Upper Palaeolithic period would in like manner have been recorded by their Mesolithic descendants.

From this background then did early Mesolithic Man venture farther and farther afield as the ice, gradually melting under the influence of the more favourable climatic conditions, revealed to him the Preseli Hills thus providing him with the opportunity to extend the area of his hunting activities.

From the peaks of the Preseli's Foel Cwm Cerwyn, Foel Feddau and Foel Drygarn, taking with him his knowledge inherited from most ancient times, Mesolithic Man was able to establish for himself the important meaningful alignments, not the least of which was the early May sunrise to a Drygarn Fawr (147,863584* over 50 miles away in the Welsh Uplands) in direct line of sight and as yet not fully free of its icy covering. (A very clear and superlative sighting from Foel Cwm Cerwyn of Drygarn Fawr over fifty miles away as the crow flies, was recorded by a reliable source on 13.7.85.)

As the years wore on and the area of Drygarn Fawr freed itself from the all-enveloping ice, compelled probably, in the main, by the idea that he would arrive at the May sunrise which he could see so clearly coming from behind Drygarn Fawr in the far distance, Mesolithic Man set out, as an explorative hunter, and made his way in a long, slow journey fraught with dangers of all kinds along the forest covered ridge in the direct line to Drygarn Fawr.

To extend the area of his hunting exploitation purely and simply would certainly not have demanded the covering of such an arduous distance as that of some seventy miles, so some other idea there must have been which drove him to undertake such a long, dangerous and, to first appearance, unnecessary trek. Could that far distant place, whence came a sunrise of so long revered significance, have stirred in that Mesolithic heart the hopes of finding a 'Shangri-la'?

From nowhere other than Preseli then, could the earliest peoples to arrive at Drygarn Fawr have come. Bringing with them their ancient heritage of Sun alignments, they laid the earliest foundations of Drygarn Fawr as a seasonal clock and using the knowledge from their Upper Palaeolithic ancestry, they appointed alignments for early February, early May, early August and early November (see Chapter VI).

That there were any Mesolithic arrivals at that early time from areas to the east is extremely doubtful. Inaccessibility on topo-

---

* Where necessary the three figures (e.g. 147,) before the grid reference refer to the Ordinance Survey (1:50000) Map number.

graphical grounds with precipitous-sided valleys forming effective barriers combined with the thick impenetrable forests and large areas of standing water and swiftly flowing torrents, all these factors taken together made the possibility of early arrivals from the eastern side remote in the extreme — quite apart from the fact that in those early days there was no compelling need either on grounds of hunting or exploration or hopeful curiosity — which latter did exist, however, in the case of the early peoples on Preseli.

The Mesolithic period differed in length in different regions according to the varying progress that the climate made towards being the optimum for human survival and progress. Clark (1936) puts the Mesolithic period in Northern Europe as occupying nearly six millenia, a period not considered too long when taking into account the profound changes in climate and animal and plant-life that unfolded themselves while the cultural developments we are discussing were taking place.

With the great advantages of more precise placing in time-scale that the comparatively recent blessing of science, radio-carbon dating has brought, it has been shown that the Neolithic way of life had already begun in a Meso-Neolithic fringe date of 4,000 B.C. and not at about 2,000 B.C. as everyone, including Professor Piggott, without the benefit of the use of radio-carbon dating, had supposed. Burgess (1980) points out that such an immediate lengthening of time-scales meant that evidence that had previously been crammed into a few centuries had now to be spread over much longer periods. These longer periods amounted to having to count in thousands of years instead of in hundreds. A lengthy re-appraisal of time-scales which had been previously accepted had therefore to be urgently made.

It must not be thought that the phrase "the Neolithic period began before 4,000 B.C." means the existence of a distinct dividing line between the Mesolithic and the Neolithic. There would have been more developments of a Neolithic kind earlier in some areas than in others and when the Mesolithic way of life could be said to have finally died away, then it was that there had been a very long phasing-out (probably, by and large, spanning more than 1,000 years) of one culture that had outlived its usefulness and, in a gradual development of its own in an attempt to accommodate to the new demands of a different set of requirements for living, had re-presented itself in a developed and progressed form called, by Modern Man, the Neolithic way of life.

Several groups or cultures are comprised within the length of time over which the Mesolithic period lasted.

The earliest groups were the Maglemosian who, as has been found from their remarkable site at Star Carr in Yorkshire where they lived in about 7,500 B.C., resembled previous Upper Palaeolithic peoples in many ways, more than they resembled later Mesolithic communities as Lloyd and Jennifer Laing in "*The Origins of Britain*" point out. The existence of Maglemosian sites has been established in Wales in Pembrokeshire, but little research has been carried out on what is the probability of their existence in other areas close by (see chapter XIV Preseli).

As man made his way through the millenia changing his living habits in ways big and small to suit a varying and variable world around him — one culture would superimpose itself on the previous one, just as is still in the process of happening all around in the modern evolving world of the twentieth century A.D.

It is intended here to pay attention to those Mesolithic peoples in particular, who, living in this hill region of Central Wales, in spite of their primitiveness and the necessity for constant battle against the forces of nature, contrived a life-style that not only permitted them to survive but also enabled them to hand on the great gift of a very precious know-how to very many generations of mankind who came after.

CHAPTER II

The world about them

When we think of the appearance of the mountains in this central part of Wales (see Fig. 18) we see them in our mind's eye as rounded in outline (with the few outstanding exceptions such as Pumlumon and the Brecon Beacons) and deeply intersected by narrow steep-sided valleys. A bare-hilled rather barren region this, sometimes boggy, sometimes aridly tussocky and suitably inhabited by sheep, crows, larks and the rare evasive red kite.

To the Mesolithic people of those times, that was not quite the picture. There was far more water about for one thing, and far more trees about for another.

The waters from the melting ice sheets of the Ice Age were already some thousands of years along the time to their final clearance, but much of what are to us valleys, as we know them today, were still the beds of lakes, and under water which had been dammed back by various natural obstructions lower down towards the sea. Many temporary lakes (even if temporary might mean 1,000 years) were thus formed, around the edges of which these peoples fished, walked their ways and constructed their frail shelters.

The traces of where the earliest peoples lived can still be made out on what to us are slopes high up on the valley sides. This fact alone gives an indication of how close it was to the time that the melting waters of the Ice Age were still in the process of dispersing, that the earliest peoples of this period lived. No Neolithic peoples these, for the sites, where the Neolithic peoples of subsequent times, some millenia later were to live were, as yet, still under water. Professor T. R. Owen, on the basis of geological evidence, is of the opinion that

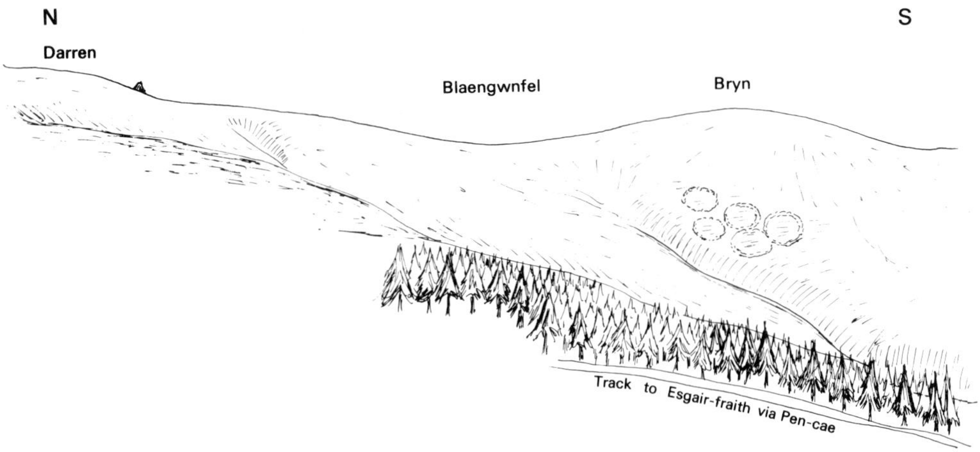

Fig. 2 *Traces of circles on Bryn at 897554.*

these sites high up on the valley sides would belong to about 8,000 - 7,000 B.C. It is very thought-provoking to see the faint traces of these primitive peoples' habitation. Some can be distinguished at a location near to 897554 on Bryn (see Fig. 2). The photograph, from which the sketch was made, was taken from the track to Esgair fraith from Pen Cae — well below in the valley, and under water at the time that the owners of the circles were doing their hunting and fishing. These circles would have been near the edge of the water level of a temporarily naturally dammed lake of melt-water and this explains the reason why they appear now to be so high up in the air.

The height of the tracks and paths of these peoples on the mountain sides is yet another witness of the date in time that the earliest peoples of our study lived. These ancient tracks are visible even today skirting what was nothing other than a lake-water level (see 147,881712 for such an area). An Ancient Road comes over from Llofftyddgleision (948718) through Pont-ar-Elan and makes for Carn Ricet and so on westward to Teifi Pools (785675). At 885713, it is skirting well above today's valley floor making for the col on the way to Carn Ricet. At this approximate point, another old track can be seen leaving it to the left, and skirting, what is today the neck of the valley, maintaining its height, (about 445m) it makes its way over and past Esgair cywion. The map shows no trace of this but a walk to see it is well worthwhile.

In this way, evidence gathers of human activity occurring in this area comparatively early and possibly almost as soon as the receding melt-waters permitted of human penetration into these particular regions. This could have been as early even as 9,000 B.C. which period forms part of what is known as the Windermere Interstadial.

This Interstadial was a period from about 11,000 to 9,000 B.C. when a considerable improvement took place in the climate which helped gradually to transform over a period of three to four thousand years an inhospitable snow and ice-covered tundra into a land of forests, trees, bushy vegetation and lakes that could be fished. By about 8,000 B.C. the climate had become rapidly more favourable to human existence and by 5,000 B.C. (the climatic optimum) it was better than it is today. It has been reliably assessed that at that time, the annual mean temperature was about 3°C higher than now — which difference had considerably greater influence on the expansion of areas of vegetal growth than 3°C might at first suggest.

This leads to the second big difference in the appearance of the landscape as seen by the peoples of between 8,000 and 5,000 B.C. from that seen by us in the twentieth century.

The bare hills which present themselves to the gaze of Modern Man were not at all so in those times. Through several thousands of years, deciduous forests established themselves, continuously expanding in area. Later people of those times, standing on the highest hill-tops, would have been looking over a vast expanse of waving tree-tops spreading as far as the eye could see.

Nor was this expansion in area of vegetation the sole result of the improvement in climate — there was also a multiplication in the variety of trees that grew. Oak, ash, hazel, birch and alder — all made their appearance — as similarly did the herds of warmth-loving animals such as deer, which displaced the reindeer and arctic fox of the tundra.

CHAPTER III

Making the best of a Life without Social Security

It must not be thought however that now everything had turned into a paradise. Man in the period between 8,000 and 4,000 B.C. was a hunter, a gatherer. Not all were hunters. Some were fishers. And with that, Man has, in many areas in which he existed, been more or less dismissed.

A significant and in depth study of the Food-gathering peoples of continental Northen Europe during the Early post-glacial period has been made by J. G. D. Clark (1936) in his book "*The Mesolithic Settlement of Northern Europe*". In this, Clark feels it as well to point out explicitly that his work is meant to appy only to that part of Northern Europe selected for his special study. Clark divides this post-glacial period into three parts making their division coincide with the periods of climatic and forest development. Chronologically, therefore, starting with —

Period I from about 8,000-7,000 B.C.	the retreat of the ice sheets
Period II from about 7,000-5,000 B.C.	with its climatic and vegetal readjustment produced by this withdrawal.
Period III from about 5,000-3,000 B.C.	with its continued growth and spread of forest and animal life in consequence of the changing habitat.

It is as well to bear in mind, however, that while Clark chooses to limit his remarks purely to his area of study, the fact remains that the ancient peoples of the present work were contemporaries in a very large part if not in the entirety of the periods mentioned. Their enviroment in general and their daily needs in particular could not have been too dissimilar allowing for the difference in type of terrain inhabited.

While many artefacts have been found which the Mesolithic peoples of Northern Europe left behind and of which Clark's work gives innumerable examples, the fact that up to the present there is almost nothing similar bearing witness to the life-style of their contemporaries in these hill regions, would indicate more that nothing has ever been seriously looked for rather than that there is nothing to be found. It is interesting to note that before the Welsh Water Authority began the construction of the Brenig reservoir in Clwyd in 1973-76, consideration was given to full exploration of archaeological sites known to exist in the Fechan valley. Several of the many sites investigated yielded invaluable information on the history of early Man at Llyn Brenig. A Mesolithic camp was found near Boncyn Arian — to name only one example — where excavators uncovered "shallow scoops in the ground containing the ashes of fires and surrounded by flint tools and waste flakes". The charcoal dates back in all probability to almost 6,000 B.C. The Stone Age hunters and food gatherers, who were the makers of these tools, were probably the first men to enter the valley after the Ice Age. (See "*Man at Brenig*" produced by the Welsh Water Authority).

It has with truth been said that any body of people existing together — even if only as a family unit — are a society, and a society has to have an organised modus vivendi. This modus vivendi is born from the needs of that body of people (often being imposed on them by the purest chance of place of birth) however limited the needs might seem. The aspect that is much overlooked is that their way of life represents the aspirations of the human spirit, and in the society of this period, the main aspiration of the human spirit was to survive.

While being wanderers, hunters, fishers, and berry gatherers, pursuing a nomadic way of life, in order to survive the day, these men had to be far from aimless wanderers and nomads. Aimless wandering meant certain death whether from starvation or from other hunters in the shape of wild animals. Wolves, wild boars and snakes abounded and were a constant threat to the unwary. The wanderings of these people had to be tinged therefore, with wariness and caution — constantly trying as much as possible to turn the unknown into the known.

Short in life (most were dead by 30 and the rest by 45) these earliest peoples tended to live in a group consisting of just the one family unit. The Laings in "*Origins of Britain*" say that evidence from Star Carr in Yorkshire suggests that in 7,500 B.C. the total population of the British Isles did not exceed 10,000. One can well imagine, therefore, that in the earliest days of our period there would not have been many family units in the Welsh Upland area, and the total population figure of 10,000 would have remained much at that level for some time, for the reason that the natural environment of any specific region was unable to give subsistence to anything other than small social groups over a comparatively short space of time, And, of course, lives would be lost as a result of hazards such as forest fires, flash floods, wild animals and exterminating epidemics.

Although large areas of forest would have been uninhabited, the family unit still tended to circulate within much the same general area of yearly exploitation, while having seasonal sites within its encompassment. This was for the reason that the locating of food was simpler in a familiar area than in a strange one. Time and energy would not be wasted in vain searching for berries and nuts in the appropriate season and, through experience of the area, there would be an awareness of the herd movements, where the herd drank in the evenings, in which direction it moved in the hours of darkness and where it could be expected to be found at daybreak day by day throughout the hunting season. Man had to run after his dinner and the shorter and less exhausting the run, then so much nearer was Man to survival.

As Man moved around his usual territory in hunting or in transferring to a seasonal site, he had to be sure that he was going in the right direction. It was easy to lose the way in the endless forest and hills especially as ordinarily recognisable mountain features could become swiftly hidden in cloud and mist.

Direction posts had, therefore, to be devised — whether it be in the shape of Standing Stones or cairn-like piles of stones placed in prominently visible positions. These are to be found today in profusion, although it has to be remembered that all such are not of necessity wayfaring sign-posts. Through the thousands of years of this twilight era, there were those that were erected that meant a different thing to the different people who placed them there at different times. For those peoples who erected them as directional signposts, they meant the difference between living to the morrow or dying miserably perhaps in the darkness through being killed by animals of the wild, or lost in the forests and dying from fatigue or exposure.

When natural catastrophe such as flood or forest-fire spread devastation, then of course a new hunting area had to be explored by those who survived both the catastrophe and its rather more threatening-to-survival aftermath. Herds would have perished in the flames too and this added yet another threat to an already sufficiently difficult life. The small human unit would have been deprived of an important source of food and for a considerable time at that, until herd survivors, if any, had built up numbers again — which could of course, well take years.

While varying in intelligence from individual to individual as does Modern Man today, even the least well endowed, brainwise, in the early post-glacial period would have realised that ripe nuts and edible berries were available at only one time of the year and not all the year round, just as he also realised that the herds calved at only one time of the year and not all the year around and that young edible leaves and buds appeared anew and grew on trees at only one time of the year and not all the year round.

That these measurably regular happenings were connected with the comings and goings of the Sun and were fully dependent on its appearance and disappearance was realised even in Upper Palaeolithic times as is evident from the conclusions of Dr Margaret Murray to be discussed in Chapter V. The presence of warm Sun produced buds and leaves and in due course, there would be nuts and berries. When there was little Sun, there was no growth and that was Winter.

With experience and in the rhythm of their lives, these peoples of the post-glacial period came therefore to realise that a more sure tracking down of available food would be secured if they noted where the Sun was in the sky — and could be expected to rise at times relevant to plant growth. It would be necessary to be able to recognise the point on the horizon of the Sun's rising and of its setting in order to relate it to the season — but what was of even greater importance, was, that through experience, being able to *foretell* this regular happening, they would then be able to make the necessary arrangements in their lives in advance. Hunters on high ground exploiting an area appropriate to the season had to be able to tell when the time was coming for them to have to withdraw and prepare to return to lower ground where the berries and nuts of Autumn would soon be ripe. That such a fore-knowledge was a necessity to survival by Man, the hunter, long before he became Man, the farmer of the Neolithic period, is easy to appreciate.

Of all systems required to be used by these ancient peoples in their fight for survival, the necessity for the establishment of some organised system by which track could be kept of the Sun's path would seem to have exceeded all others in urgency and importance and his actual establishment of what might be called seasonal clocks where the Sun's movements could be evaluated and foretold must rank, bearing in mind his primitive state, as one of Man's most outstanding achievements of all time.

CHAPTER IV

How they set about getting it or Developments of a Mesolithic kind

Drygarn Fawr (147,863584)

Drygarn Fawr is not one of the easiest hill-tops to reach — largely owing to the tussocky nature of the ground for the last mile or so. It is best to approach it from Llanwrtyd Wells (147,878467) taking the road to Abergwesyn where one turns left and proceeds as far as Llanerch-Yrfa at 836556 where the car must be left. There is a path marked on the Survey Map but reafforestation has largely obliterated it up to the forest fence. A way can be found, however, up through the trees by any who still have some of the spirit of the old British explorers left in them. There are locked gates in the forest fence by climbing which the mountain land is reached. The mountain land is not common land and is privately owned. If you cannot find the path, proceed from the forest fence in a general north-easterly direction gradually ascending in a crosswise path the rising ground on your left — and when you have reached 856574 approximately, Drygarn Fawr will be in sight — as will, below you, the path that you failed to find. It is best to cross the stream where the path crosses it as you may otherwise find yourself in very unpleasant bog. Allow a good 3 hours from Llanerch-Yrfa.

The most outstanding promontory of this area of the Welsh hills is doubtless that of Drygarn Fawr (641m) O.S. 147,863584. Centrally placed as Drygarn Fawr is, in nothing other than matriarchal majesty, the panorama from the crest will, on a clear summer's day, encompass the Stretton Hills of Shropshire, the Carneddau near Builth, Hay Bluff, the Black Mountains, the Brecon Beacons, the Eppynt range, the Carmarthen Fannau, Preseli, Pumlumon and beyond. Just as Modern Man gazes in wonder and amazement at the 360° panorama about him, so must his ancestors of so many thousands of years before Christ have gazed about them.

These geographical locations named are for the most part on the most distant horizons with innumerable heap upon heap of promontories and ridges intervening. These most distant horizons are frequently very difficult to distinguish, but Ancient Man would have been in no doubt, after sufficient observation had been carried out over a considerable length of time, as to the exact point at which the Sun appeared and disappeared relative to the seasonal happenings around him. Knowing the difficulty with which the actual distant horizon could be distinguished, he chose intervening marker promontories to help him remember the exact point on the horizon beyond. These were usually the nearest available and the most easily identifiable. In the case of Drygarn Fawr, it was truly wondrous how conveniently and accurately intervening hilltops presented themselves as markers of astronomic significance. That, to Modern Man, in all his sophistication in this field, inaccuracies of varying degree are apparent, is only so in terms of sophisticated modern scientific measurement. To Ancient Man the system worked efficiently enough in the light of his needs. Even so, when it is considered that Ancient Man had only his eyes and Nature's horizon and his own power of memory as instruments, then his degree of accuracy is truly astounding. It would be moreover rash to conclude that Ancient Man, however primitive he might have been, did not introduce some degree of selectivity in his choice of intervening aligning promontories. It is evident that he had considerations of allowable and unallowable approximation in the light of even his primitive standards.

Accepted aligning promontories all bear signs of human activity on their crests and slopes. Once a hilltop had been found, therefore, that was in an alignment of some astronomic significance, a little expedition must have set out to mark the fact. Some were found to be of greater imporance than others. The importance of Corngafallt (147,943644), for example, is obviously outstanding. It not only provided an alignment from Drygarn Fawr, but also had its own aligning hilltops of astronomic importance in other directions. Such a hilltop bears signs of human activity and human habitation over a considerable length of time (see Chapter VII). It must not be overlooked that it is possible that being in the path of an alignment might have invested the particular hilltop with a certain strange power in the eyes of the watchers of the horizons. This is nothing other than pure surmise of course, but the fact remains that, hilltops *not* in the path of an alignment seem to have escaped human attention and those with only one alignment do not seem to have received the attention that was paid to an aligning hill-top with *several* additional alignments of its own or that paid to an aligning hill-top on an alignment of *particular importance*.

If it were possible, therefore, to fly high enough in some hovering type of aircraft to take in the whole of the area covered by these solar observatories of the Welsh hill area and, at the same time be low enough to be able to examine the crests of all the relative aligning hill-tops — then there below would lie revealed all the secrets of Ancient Man and his solar observatories.

While Modern Man attaches importance as to what moment in time, as the Sun's orb rises progressively over the horizon, it can be considered to be the moment of sun-rise, such sophisticated differentiation did not matter much to these earliest of peoples. What mattered was that they should know in advance to what part of the horizon they should turn their expectant gaze — and it was the *guide-line* of the position of the intervening marker promontory that permitted them to do this — a feat impossible over a *flat* landscape to a distant horizon where the lines of sight would have been broad and diffuse.

Of such materials then, as convenient hill-tops, did Ancient Man construct his early solar observatories, his seasonal clocks. After a long period of observation over many years, it would be confirmed to the watchers on Drygarn Fawr that to turn their expectant gaze in the general direction of what is known to Modern Man as Corngafallt (147,943644) and to see the sun rise from a point somewhere behind that, then the Midsummer sunrise was taking place (See Fig. 5). In addition, it would be appreciated that this was the longest day in the middle of the warmest season and that thereafter the Sun would be right-handedly proceeding on a course of retreat back to its Autumn and Winter quarters. When the Sun sank behind Freni Fawr, (145,199352) to Ancient Man this would mean that Winter had begun (early November) and to Modern Man this would be a reading of 250° which sounds totally uninteresting!

A day of rejoicing would come when the Sun once again would rise from behind the direction of Darren (110° to Modern Man) thus marking the end of Winter and meaning that it was starting on its slow journey of return, bringing with it Spring, and the renewal of life. That it would take a considerable length of time to build up the total picture for a solar observatory is very obvious, but this was such a vital piece of knowledge that these ancient peoples would have given considerable priority in the acquiring of it.

As one familiarises oneself with these various centres of solar observation and the complexes arising therefrom, one wonders more and more at the ingenuity of these primitive peoples. Knowing nothing of compass points, leave alone refraction, parallax, azimuths, zenith distance and the Haversine Formula of spherical trigonometry, primitive Man, our Ancestor, designed himself a seasonal clock. And if a seasonal clock is one of acceptable astronomic alignments, then it has to be accepted by this very fact that it is of very real astronomic significance. And this apparatus, if it can be called that, was as wonderful to Ancient Man as the inner sanctum of Mission Control, Houston is to his modern counterpart.

While these earliest of all time calculators depended heavily on the whims of Mother Nature to provide conveniently aligning hill-tops, the day was as yet far distant when Man would abandon his dependence on the fortuitous presence of these promontories in favour of contriving for himself a D-I-Y kit — a kit using stones, which he could move about as on a draft board, placing them with accuracy by hand in positions of sun and moon alignments that his long experience in various sites in the mountains had taught him.

By the time he was ready to do this, his advancement in sophistication from the simple Drygarn Fawr and its neighbouring complexes was to be very evident and was to show Modern Man that that final magnum opus of Stonehenge when it finally came, was in its general principle no innovation whatseover — having had as forerunners all the solar observatories in the Welsh upland areas from those of the Drygarn Fawr complex and the Carn Ricet and Carneddau areas to those of Preseli.

CHAPTER V

Classification and explanation of critical directions referred to in the text

Part 1. The Sun alignments and a Memo.
Part 2. The Moon alignments and a Memo.

The directional theme is the tracking by Ancient Man of the movement of the Sun and the Moon. For practical purposes, however, corresponding directional values recognised by Modern Man have to be used as e.g. 050°, 070° or 310°. 050° (that is 50° East of True North) is generally used to indicate midsummer sunrise and all other readings are on this generally accepted modern day basis.

Part 1 *The Sun Alignments* (see Fig. 3)

End of June	Midsummer sunrise. Longest day shortest night. From now the Sun would go back whence it came to give shorter days and longer darkness.	050°
Early May	sunrise. The Sun is approaching — buds and leaves grow again and the hunted herds have their young. Spring-early Summer.	070°
Early August	sunrise — the Sun is still going back whence it came.	070°
Late March Late September	sunrise. Equinox (equal length of day and night). Autumn starts with its harvest of berries and nuts.	090°

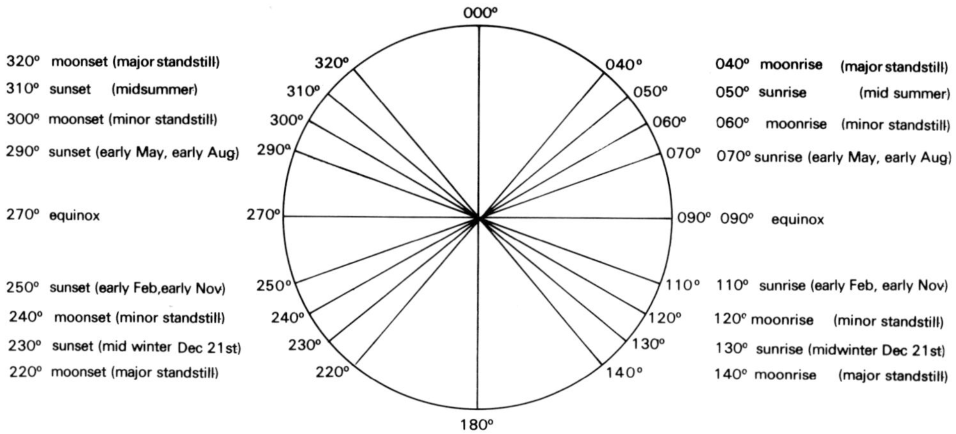

Fig. 3. *Sun and Moon alignments of the compass as used in this book.*

Early February	sunrise — the Sun is approaching — winter is ending.	110°
Early November	sunrise — the Sun is still going back whence it came and the true cold and barrenness of winter is near.	110°
End December	sunrise — this is now the middle of Winter.	130°
End December	sunset — shortest day, longest night. After this the Sun would start to return and spirits rose.	230°
Early February	sunset confirmed that Winter was ending and that Spring would soon be coming.	250°
Early November	sunset — the Sun was now well on its way to its winter quarters and Winter had almost arrived.	250°
Late March Late November	sunset. Equinox (equal length of day and night).	270°
Early May	sunset — evidence now of Spring and its accompanying renewal of vegetational and animal life.	290°
Early August	sunset — Autumn — berries, ripening fruits, nuts.	290°
End June	sunset — midsummer — longest day, shortest night, henceforth days would shorten, hours of darkness lengthen.	310°

Memo.

The meaningful Sun alignments existing to mark early February, early May, early August and early November, (such dates occurring as they do at an approximate half-way stage between the solstices and the equinoxes) played a significant part in a fully efficient solar calendar.

Dr Margaret Murray in "*The Witch cult in Western Europe*" (1962) states quite categorically that "such a division belongs to a very early period before agriculture. It has no connection with sowing or reaping," she says, and goes on, "It ignores the solstices and the equinoxes but it marks the opening of the two breeding seasons of animals, the wild and the domesticated. It therefore belongs to the hunting and pastoral periods and is in itself an indication of the extreme primitiveness of the cult and points to a very early origin reaching back to the Upper Palaeolithic period (Old Stone Age)."

In the development of time, February, May, August and November turn up as great festivals of the Celtic tribes. In the opinion of Burl (1976) the dates of the four annual Celtic feasts — Imbolc, Beltane, Lughnasa and Samhim could be events based upon these very anciently revered dates. Imbolc was the celebration in early February when the birth of young to the herds and the coming of milk was a period of great rejoicing and was to all intents and purposes the beginning of a new year. Samhim, on the other hand, in early November was when the decline of the Sun produced great fear that it might not return.

Burl in "*The Stone Circles of the British Isles*" (1976) expresses the opinion that, "for very early peoples, the rising and the setting of the Sun on such dates had perhaps even more profoundly important meanings — meanings that governed their wanderings and gave information on food availability as the Seasons advanced and retreated".

From the time so many thousands of years ago when they were first noted by Man, traces of the existence of these dates have survived even to the present day right down by way of the medieval witch cults of Western Europe when their celebration was widespread. In Modern-day Wales, Calan Gaeaf is celebrated in November as the onset of Winter, Calan Mai in early May (surviving as dancing around the Maypole in countries as far apart as England and Czechoslovakia) and the Calan of early February, which was considered as the true New Year and still is, in some parts of Wales, as in the Gwaun Valley of Pembrokeshire. Here this celebration takes place in modern-day terms on January 13th but it cannot fail to be noted that this date falls within the important period of the year when the Sun in its risings is approaching a 110° alignment, marking the end of Winter and bringing the expectation of a renewal of life.

Wood, in "*Sun, Moon and Standing Stones*" (p. 169) refers to the presence of these self-same alignments in Stonehenge commenting that "they serve as a possible basis for a solar calendar".

That meaningful alignments such as these are to be found in the seasonal clocks of the Welsh hills does, of course, indicate that something in the nature of a seasonal clock must have existed in Palaeolithic times — which leads to the important conclusion that their existence in the seasonal clocks of the Mesolithic period amounted to a plain statement by the peoples, who cherished such alignments, of a very ancient heritage from far, far back before their own time.

Part 2. The Moon alignments referred to. (see Fig. 3)

Moonrise in northerly major standstill	040°
Moonrise in northerly minor standstill	060°
Moonrise in southerly minor standstill	120°
Moonrise in southerly major standstill	140°
Moonset in southerly major standstill	220°
Moonset in southerly minor standstill	240°
Moonset in northerly minor standstill	300°
Moonset in northerly major standstill	320°

In the course of just twenty-eight days, the moonrise and moonset directions swing in two arcs, respectively, from a northerly point (or "standstill") to a southerly point (or "standstill"). The swing of this arc varies from a maximum (major) angle to a minimum (minor) one. It takes that arc 9.3 years to change from a major standstill swing to a minor one, or 18.6 years to complete a full cycle. Fig. 3 illustrates the directions of major and minor moonrise and moonset standstills.

Memo.

The Moon does in one month what the Sun does in one year — passing from New Moon to Full Moon in about 14-15 days and then decreasing in the next 14-15 days to restart the cycle as a New Moon.

That the study of the Moon seems to have begun later than the study of the Sun is probably explained by the fact that the Sun was for obvious reasons more vital in Ancient Man's need to survive than was the Moon. This does not mean to say that Ancient Man totally disregarded the activities of the Moon in the earliest times. Its alternation between New Moon and Full Moon and Full Moon to New Moon was too conspicuous by its very frequency nor was there any necessity for any alignments in this earliest of simple conceptions. It is indeed quite probable that the Moon's cycle did to some extent provide a rough and ready time gauge for man in his most primitive times as well as introducing into his life a certain mystique as he looked around him at his normally familiar surroundings transformed by the cold eerie light of the Full Moon.

As he progressed, the Moon was to be used for measuring shorter periods of time than that of a season's length. The number of moons would give the number of what we call months. And later still, more in a spirit of scientific discovery and search for knowledge rather than in one of survival, he was to be wanting to foretell the Moon's eclipses.

This in itself indicates the great length of time that those seasonal clocks spanned in their build up, as well as showing a great step forward in the progression of thought of which Man had by that time become capable.

CHAPTER VI

Drygarn Fawr as a seasonal clock

(Figs. 4 & 5)

Geological Background

The outstanding ridge of Drygarn Fawr is formed of massive beds of pebbly grits and conglomerates (the Drygarn Conglomerate) occurring within a thick Lower Silurian sequence of blue and grey shales and mudstones. The conglomerates occur particularly on the southern flank of the Rhiwnant Anticline and were probably deposited as an extensive submarine turbidity fan in early Silurian times. Today the steeply-dipping conglomerate beds form marked ridges on the southern slopes of Drygarn Fawr with the intervening shaley layers being marked by hollows or terraces (see Fig. 6a).

(I am indebted to Professor T. R. Owen for the foregoing geological background.)

1. It requires to be emphasised that many of the points named (see Fig. 4) were merely pointers to a direction on a far distant horizon not specific horizonal points in themselves, e.g. Esgair Fraith (894566) is a direction pointer to Hay Bluff, Pen-y-Gyrnos (777513) is a direction pointer to Cairns placed at a point on the horizon behind and beyond on Pysgotwr, often hidden in cloud or mist and particularly hard to distinguish specifically, in the tumultuous heapings of peaks crowding in the area.

2. That these directional intermediates were of the utmost importance is placed beyond doubt when these sites are themselves

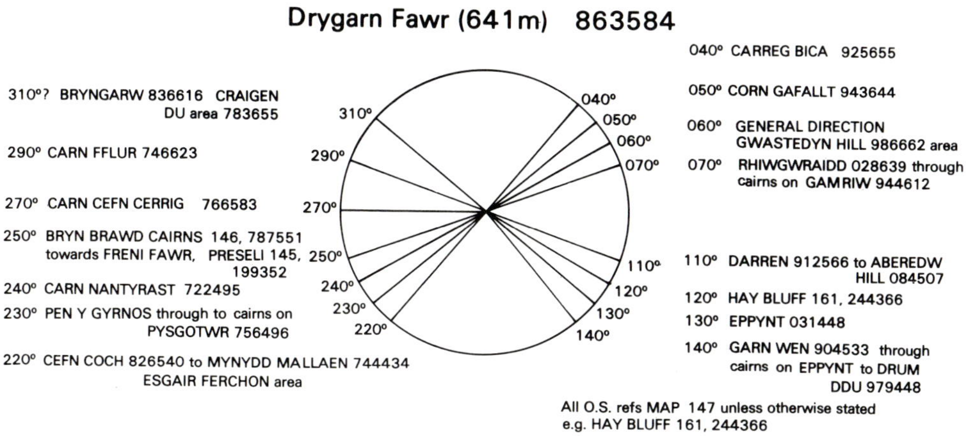

Fig. 4. *The Drygarn Fawr alignments.*

examined. Ample evidence of human activity can be seen in spite of many factors that could have — and indeed in many cases have — erased much sign. Wind and weather and the passing of thousands of years, subsequent human activity such as the habit of one culture overlaying a previous culture, agriculture, the construction of 'trig' points and re-afforestation, these have all contributed to a certain faintening of the traces of the original inhabitants.

3. It is an interesting fact that while many promontories exist on which there is no sign of human activity of any kind neither do they furnish any alignments of astronomic significance.

4. The diagram (Fig. 4) shows Drygarn Fawr as complete a seasonal clock as it ever probably was in its finality. It is impossible to guess how long it took to achieve this or whether even the first peoples there ever envisaged originally more than a few of the alignments most urgent to their needs — such as possibly the mid-Summer sun-rise and mid-Winter sun-set and the approach of Spring. That so many marker promontories existed from Drygarn Fawr for so many important directions was a kind quirk of Nature not dispensed so freely in other locations as the reader will see from subsequent diagrams.

5. The sites to which meaningful alignments exist from Drygarn Fawr would obviously have provided what Modern Man, at least, would recognise as nothing other than a reciprocal back to Drygarn Fawr. Just as there is an 050° from Drygarn Fawr to Corngafallt so is

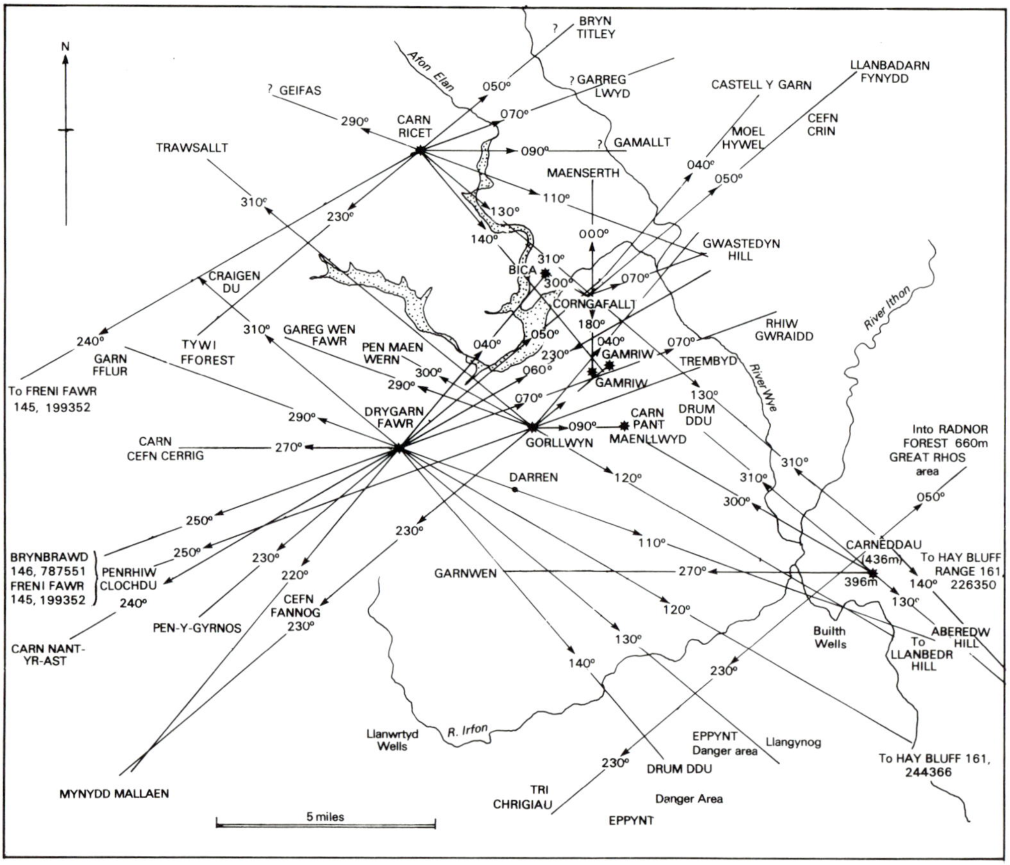

Fig. 5. *Geographical representation of the Drygarn complex. An overall picture of the Drygarn Fawr complex. It never existed in such entirety at any one time but it existed from some of the centres some of the time – often spasmodically and over a period spanning some 6,000 years or so. There were never many people at any one time but there were always a small number of people over a very long period of time.*

there a 230° (mid-Winter sunset) back from Corngafallt towards Drygarn Fawr (see Fig. 6b). It can be imagined how this invested Drygarn Fawr with quite an amount of reverence and no doubt a certain amount of fear as the forbidding black outline of Drygarn Fawr was etched against the dying rays of the setting sun at a time of the year of such outstanding importance.

6. While there are a considerable number of observatory sites in the area, it is probable that only some were ever in use at any one time and it is of the highest improbability that all were ever in use all at the same time. As has already been pointed out, in 7,500 B.C. there were

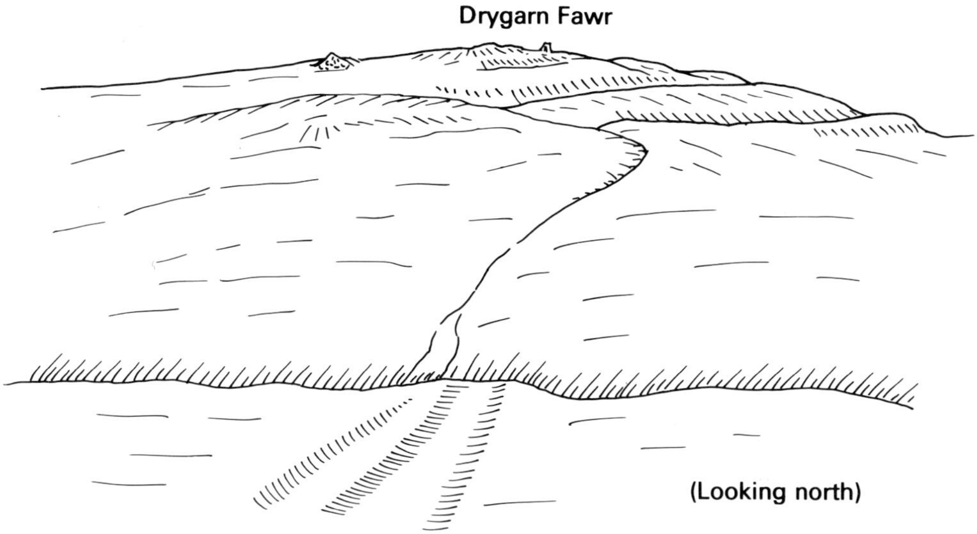

Fig. 6. *(a) Drygarn Fawr from the South. Taken from the path where it crosses the stream (see text).*

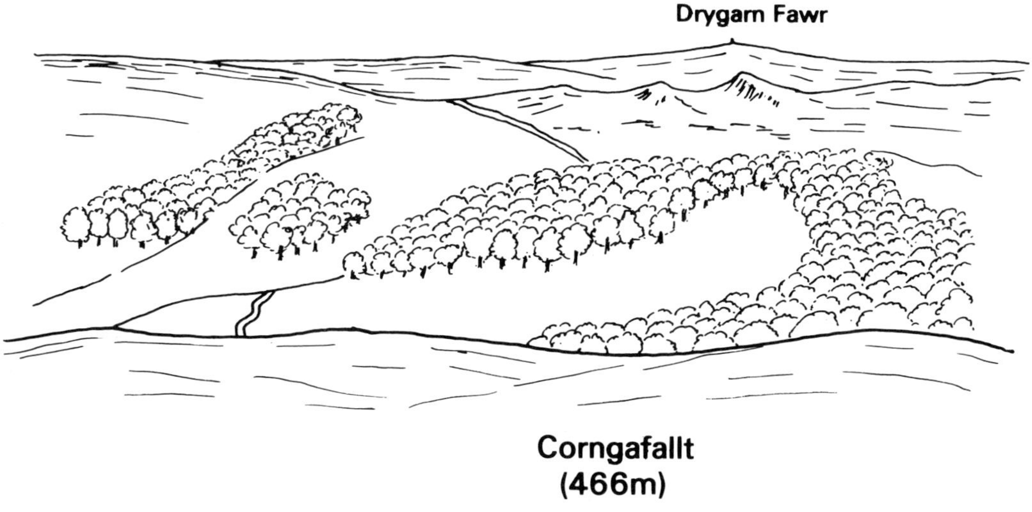

(b) Drygarn Fawr from Corngafallt. This is a mid-winter sunset alignment.

probably only about 10,000 people in the whole of Britain with no great likelihood of any significant growth in population for some considerable time. There were, therefore, great lapses of time between the establishing, lapsing and re-establishing of sites while human units died out or moved on. It would seem, however, that Drygarn Fawr itself was a hub used by many peoples and more peoples for a greater length of time than most other observatories owing to its outstanding advantages in having a proliferation of astronomic alignment markers.

Plate 1: *Drygarn Fawr from a point near where the path from Llanerch Yrfa crosses the stream.*

Plate 2: *Drygarn Fawr – marked X on the horizon from Esgair-Nant-y-brain 147,835508.*

Plate 3: *Drygarn Fawr – marked X on the horizon from Gwastedyn.*

7. This upland area of flat-topped hills and valleys around Drygarn Fawr was even in earliest times — i.e. in very early post-glacial times — better than many other areas in terms of human usefulness, as it was not as severely glacially scarred and dissected as areas farther North, such as Pumlumon and Snowdonia. The melting ice, streaming down from the higher ground, deepened valleys which furnished ways across it. Water courses later to evolve into the Wye, the Severn and the Ithon to the East — certainly not in the form that they are today — were later to carve routes which facilitated passage however limited from the East and South, while to the West and South West were those water courses which were to evolve into the present-day Teifi and Tywi.

8. The flat top area of Drygarn Fawr itself has been the unfortunate victim of considerable human interference — both in the construction of an O.S. triangulation point and in the modern construction of a stone beacon. Unfortunately therefore, little of interest is discernible. There are, however, several areas of considerable interest in the close vicinity.

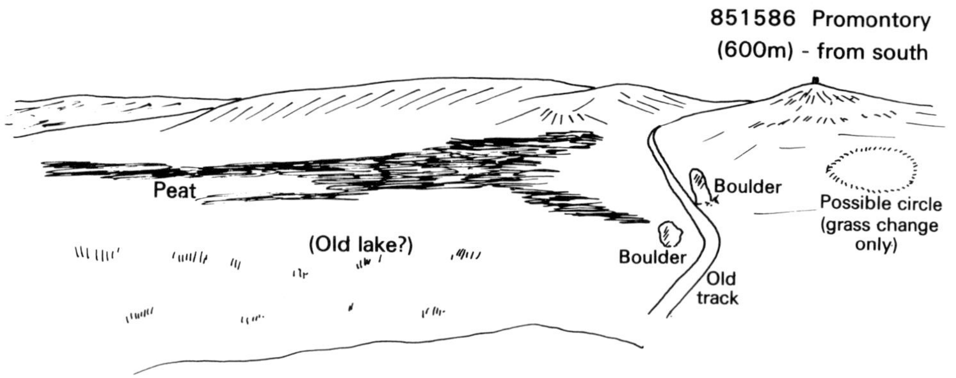

Fig. 7. *(a) Visible remains of ancient track marked by large stones and skirting the bed of an old lake at 851586 (near Drygarn Fawr).*

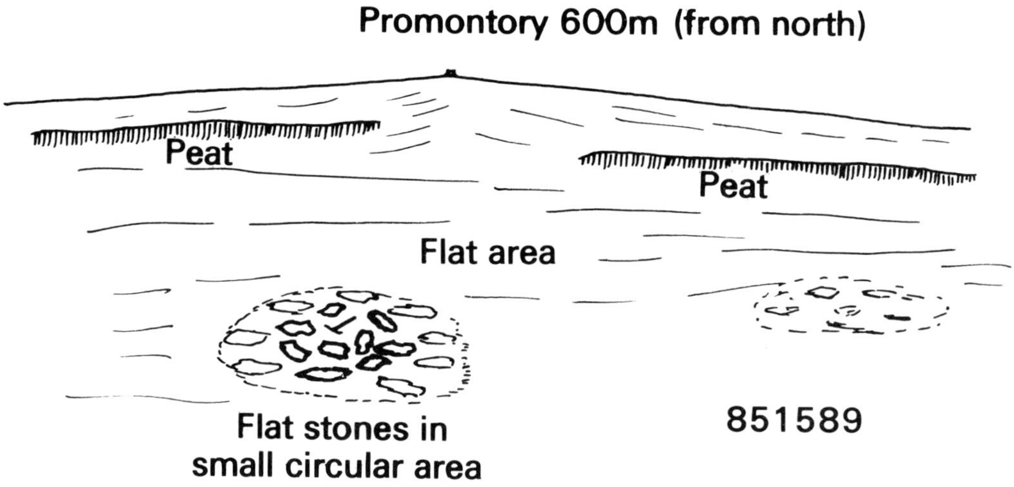

(b) Round circles of flat paving stones near 851589 (near Drygarn Fawr).

9. South of a 600m promontory (147,851586) in the region of 849584 are the visible remains of a very ancient track (see Fig. 7a) winding brokenly for about ¼ mile towards promontory 851586. Its winding course is marked by large stones and it skirts the bed of an old lake before climbing to the promontory. On the south slope of the promontory, there are what appear to be the possible remains of a circle — discernible from the distinct change of colour in the grass and vegetation. The promontory itself has what is a compact, very tidy, obviously man-made mound and ditch top, putting one in mind of the successful entry in a sand-castle building competition.

10. North of this promontory towards Drum yr Eira after an endless clamber over weather-eroded 4ft and 5ft high ridges of peat in the region of 851589, there is a large interesting flat area of some ¾ acre of dried out peat bog. Laid out in this are several round circles of what appear to be small flat paving-stones of size in proportion to the hammer depicted in the drawing (Fig. 7b). It is visually obvious that a peat layer five to six feet thick was deposited which, in the course of time covered these paved circles. It would seem that in a further subsequent process of natural erosion, this peat layer became removed, exposing these paved circles to view once again. Professor T. R. Owen is of the considered opinion that the explanation of the paved circles is not a geological one. For this reason it has to be concluded that they are man-made for some purpose yet to be discovered and of importance to those who constructed them. J. G. Evans in "*The Environment of Early Man in the British Isles*" (1975) says that with the increase in rainfall at the beginning of the Atlantic period, it was these same areas (encampments high up in the mountains) which were most affected by flooding or growth of peat and tufa, and the Mesolithic sites were buried under lake clays, peat and tufa. This would seem to have happened in the case of camp number 851589! Evans goes on to say that in the dry climate at the end of the Boreal period, lake levels fell and the surfaces of marsh dried out. The encampments of these Mesolithic peoples can be seen high up. This would specifically seem to be the case in camp 851589, while the drying climate with its consequent fall in lake levels can be the explanation for the position of camp 897554 on Bryn (referred to in Chapter II), (Fig. 2).

Evans is of the opinion that the sites to which he refers in his book represent only one of a number occupied by a family of hunters. Throughout the year a range of territories may have been exploited, each presenting different economic and environmental circumstances appropriate to the seasons of the year. This was in his opinion the extension of a tradition of Upper Palaeolithic hunting that died hard. To these peoples, with this way of life, a knowledge of the seasons in advance was necessary. They had to know in advance in order to make a start on moving back to the appropriate quarters for the season. This is so in the opinion of Evans (1975), and, added to a similar opinion expressed by Burl (1976), together with the evidence accumulated here, would seem to put the remarks well outside any possibility of speculation. Great was the need for the knowledge and pre-knowledge that a seasonal clock would provide in those very areas where such hunters existed. And so much more reason therefore for the comparative proliferation of these seasonal clock sites, however limited their information, throughout the range of territories being exploited!

11. On Esgair fraith (894566) there is a cairn and mound and ditch very much within sight and sound of Drygarn Fawr itself, and south of Esgair Fraith at 897554 (see Fig. 2) as has already been mentioned in Chapter II, the interesting group of five large grass circles — identifiable from a most positive change in type and colour of vegetation. These are high up on the hillside of Bryn (479m) indicating that at the time they were in use, they were near the edge of a lake-water level, water that had been dammed back by some natural obstruction as it made its way to the sea as melt water from the Uplands (see Chapter II).

12. While from Drygarn Fawr there is a 270° Equinox reading, a suitable 090° Equinox reading was obviously not found, but there would have been a reciprocal from Carn Cefn Cerrig which might well have been used at the appropriate time.

13. The position of Freni Fawr in Preseli in direct line of sight from Drygarn Fawr is so remarkable that it cannot pass without comment. While the Bryn Brawd cairns (146, 787551) mark Bryn Brawd as the nearest and most immediate 250° alignment for these ancient people, the pyramidical mound of Freni Fawr nestling far, far away at an unbelievable distance, must have had an eerie fascination to all who were watching and waiting for this early February and early November sunset. From the very earliest most primitive days of Drygarn Fawr as a solar observatory, the distant distinctive mound must have been a subject of wonder.

As the span of time that is under consideration is virtually four thousand years — from approximately 8,000 B.C. to approximately 4,000 B.C. — (more than double the time that separates the 20th century from the birth of Christ) it will be appreciated that the people of the early days were totally different from the people of the later periods in attitudes, abilities and sophistication. A study of the area between Drygarn Fawr and Preseli has made it apparent that travel took place between these two regions. More is written about this in Chapter I and Chapter XIII.

14. The examination of the site of the alignment for the midsummer sunset (310° Craigen Du area) is not entirely satisfactory — owing to difficulty of access. It is however hoped to investigate this site more fully in the near future.

15. In the matter of artefacts, little work has ever been carried out or enquiry been made into the existence or whereabouts of such. If, as has been shown, people lived for so long in these hills, artefacts there must be — no doubt buried under the thick layers of peat that are such a characteristic of these mountain tops.

In Vol. 78 of Archaeologica Cambrensis, 1923, p.155, reference is found to a flint factory at Abergwesyn (near to Drygarn Fawr) in a field called Dol-y-maen on a farm then, in 1923, known as Pentwyn. This field was noted for flint-stones, flint chips, flakes, arrow heads and scrapers. Those found were then considered to belong to a late period in the history of flints. In Arch. Cam. 1912 a flint factory found in Aberystwyth was described, whilst as noted in Arch. Cam., 1930, p.200, arrow heads were found in Bugeilyn in the Pumlumon area. Dr Iorwerth Peate was of the opinion that these flints were Irish in origin and could only have arrived in the area by human means. These do however belong to a period rather later than the one that is of particular interest to us, but the pity remains that no serious investigation by a knowledgeable authority has ever been carried out on what really might lie in the preserving peat and, when found, throw light on what is still a tantalisingly frustrating twilight period.

Pen-y-Gyrnos (456m) 778513
An easy walk for a change

Pen-y-Gyrnos, while it is not a solar observatory in itself and has only one alignment to it — that of an approximate 230° from Drygarn Fawr with a reciprocal from itself of a midsummer sun-rise back to Drygarn Fawr, is easily accessible, is a good walk and provides a good view of the shape of the tops of the mountains in this area for the interested but not too serious walker (about two miles of gentle upward slope). It is on private land, in a heavily sheep farmed area and the courtesy of asking permission is just as well. The farm Nantllwyd (783524) has to be passed to reach Pen-y-Gyrnos by the easiest route.

Drygarn Fawr is visible from Pen-y-Gyrnos through a cleft in the tree-clad hills — similarly in all probability, as it was in Mesolithic times. On two visits out of three by the writer Drygarn Fawr has, however, been invisible because it was shrouded in mist!

From Pen-y-Gyrnos, Penyfan in the Brecon Beacons is easy to distinguish as is Fan Gihyrych, Fan Nedd and Fan Llia. Carn Nant-yr-Ast can also be identified (another of the alignments from Drygarn Fawr as will be seen from Figs. 4 and 5) and a probable sighting of Carn Fawr (485m) 708572 far north in Caron is a pretty solid identification.

There are no signs of stones or cairns but there is a track leading directly to its highest point used now as a tractor path by the modern farmer — but it is clearly ancient and bears a great resemblance both in its width and in the way it winds to other ancient tracks to which one becomes so used in tramping these hills. There is an interesting encircling but broken mound and ditch stretching in a far-flung arc around the flanks with a gap which would seem to indicate an old path of entry. As with other easily accessible sites, it is quite obvious that much has been lost which was once there — and such remains as there are of an encircling earthwork would probably date to a much later hill fort or encampment.

CHAPTER VII

Corngafallt

(Figs. 5 & 8)

(147,943644)

"Je m'assieds en silence . . ."

"Sur le dernier sommet des monts, d'ou le regard
Dans un double horizon se répand au hasard,
Je m'assieds en silence, et laisse ma pensée
Flotter comme une mer ou la lune est bercée –"

Lamartine
"L'infini dans les Cieux"

Corngafallt is best reached by taking the B4518 from Rhaeadr to Elan Village and then proceeding along the winding track through the woods to a point near Talwrn (147, 940641) where the car has to be left if it has not already been left at Elan village. The walk is steep and scrambly but pleasant and the view becomes more dramatic with each yard gained.

Having scrambled and climbed to the top of Corngafallt, there is a sudden feeling of having walked into and intruded on the privacy of a human settlement. A human settlement thousands upon thousands of years older than Rome itself — which, of course, it is.

On the summit slopes of Corngafallt can be seen so perfectly how one culture has overlain a previous culture — not perhaps overlain exactly — but rather more how one culture set itself up literally at the side of the last culture hardly interfering with the previous site in any way. In few places as on Corngafallt is one so aware of the traces of

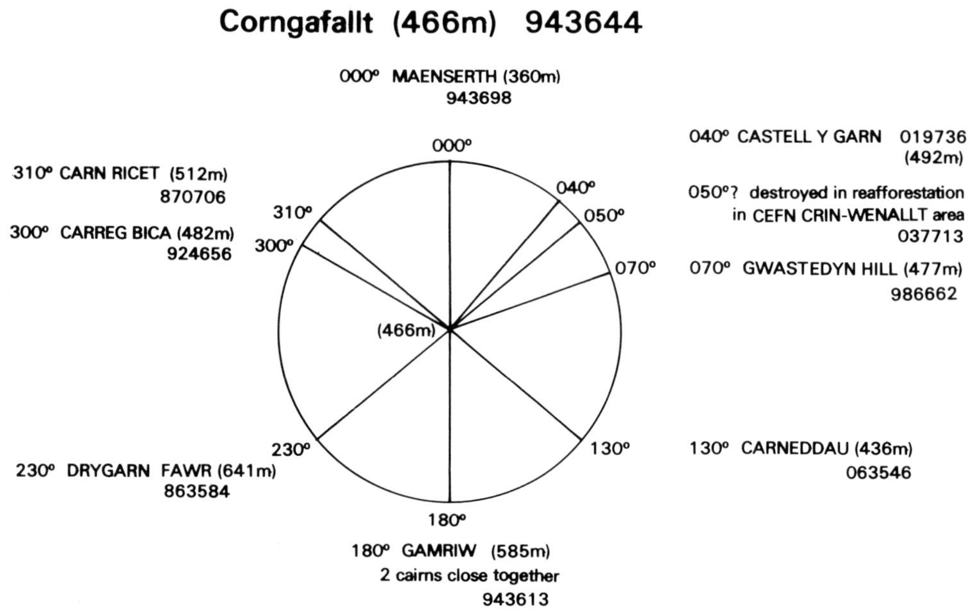

Fig. 8. *The Corngafallt alignments.*

man's passing through the millenia — and of course, for that very reason. Here is a silence and a sense of patient waiting. Here is the picture-book of the toils and partially attained ambitions of a successive train of those ancestors of ours as they went on their sad little pilgrimage through life, generation after generation over an intermittant period of what is probably four thousand years.

Here on the quite steep westerly and south-westerly facing slopes of Corngafallt, they have left in small communities the remains of their grass circles. Here too a little farther along, they have left their groups of ditch and mound circles. Farther still, they have left the remains of their stone circles (see Fig. 9a) — the stones somewhat scattered now but traceable enough. Farther along again, they have left oval-shaped elongated mounds of stones with signs of collapsed interiors and with two small flat slabs of stone still upright and about two feet apart indicating an undoubted entrance way. On a lower level are three very large and still imposing cairn-like piles of stones in direct line-of-sight to Drygarn Fawr and still somehow conveying a strong sense of purpose and significance.

On a shelf-like ledge at about 450m is the trace — still damp and boggy — of a pond or lake near which are what appear to be two

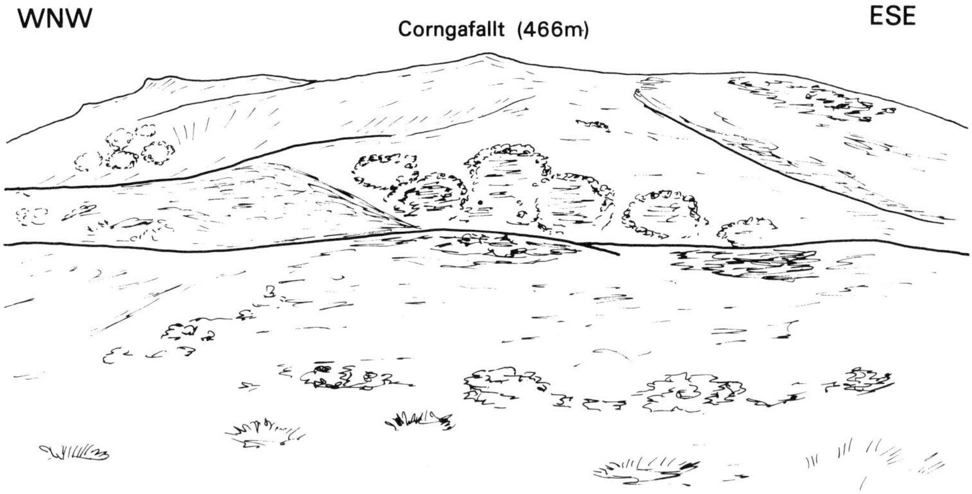

Fig. 9. *(a) Circles of stones on Corngafallt. Some circles marked only by vegetal colour change.*

(b) Standing Stones near the old lake on Corngafallt.

standing stones (Fig. 9b). The water-table previous to 4,000 B.C. was higher — so it is of no surprise that there should have been a pond or lake at this height. What is interesting is that these ancient peoples — whichever they were — those of the grass circles, those of the ditch and mound, or those of the stone circles or even those of a later period still (which is least likely) decided to make their water deeper and hold it longer by building a 'dam' (let not this word conjure up in one's astounded mind the possibility of heretofore unknown feats of Mesolithic civil engineering!). The rock of the 'dam' is varied — massive grit slab here, flaggy mudstone there, cleaved slate next and conglomerate next. For this reason, there is no question of it being a natural exposure but rather a gathering together of whatever in the way of rock could be found in the close vicinity to stop up any overflow and allow their pond to deepen a foot or so.

The main alignments from Corngafallt are shown in Fig. 8. From the solar observatory point of view, most of the alignments are Sun alignments — and important ones. There is a mid-winter sunset to Drygarn Fawr and beyond (see Fig. 6b). This would no doubt have been a scene of some drama as Drygarn Fawr would have been known as possessing several other awesome talents besides this one. For could it not itself foretell so many other Sun-happenings? There is too, a mid-summer Sun-set in the direction of Carn Ricet.

It is known that later peoples, such as the Celts, considered a sunset more important than a sunrise. To them, day merely followed night. This could have been a 'carry through' to them from the Mesolithic days which are under consideration here. The importance of this reflects itself even down to today in the Welsh language of modern times where a week is a "wythnos" (eight nights) and a fortnight is 'pythefnos' (fifteen nights). Ancient human practice dies hard.

There is a visual North-South line from Maenserth (943698) to certain markers on Gamriw (943614). Markers were placed on the promontory of Maenserth just as they were on a spot on the long ridge that Gamriw forms. There can be little doubt as to their meaning but they would hardly have been placed by the earliest peoples of the period under consideration as the question of a North and a South is not considered as having been of importance to *earliest* Man. The importance of this North-South alignment becomes more apparent, however, when one studies the more complex Preseli area (see Fig. 16. Chapter XIV).

Finally, one is left with a strong feeling that were this site in any European country other than Wales U.K., a preservation order would have been made and the site fenced off as a much revered monument to our heritage and one to be protected from any form of desecration, a site to which teachers would take buses of school-pupils to stare open-mouthed as they do in the Coliseum in Rome which is "Modern times" in time-scale compared with Corngafallt.

CHAPTER VIII

Gorllwyn

(613m) (Figs. 5 & 10)

Drygarn Fawr is clearly visible from Gorllwyn (within a good wave) but Gorllwyn like Drygarn Fawr is particularly difficult of access.

The comparatively flattish peat-ridged summit, once reached, covers a large area and consists entirely of these irregular eroded ridges of peat — varying in height from two to five feet and of length immeasurable — all of which have to be clambered over in seemingly endless succession.

The easiest way of access may be from Drygarn Fawr which is itself about a slow three-hour tussocky climb from the nearest road — so quite some length of time would have to be allowed if Gorllwyn was intended as the final goal — especially bearing in mind the return journey. This separates the Freya Starks and the men, from the boys.

Another way to reach Gorllwyn is from Marchnant 906613. The climb is long and very steep but very panoramic when one turns round to take breath but there is no possibility of avoiding the peat-ridges of the summit.

There is another way, however, from the B4358 Beulah to Newbridge-on-Wye road. Leaving this at 948532 and passing Tŷ-bach, the car has then to be left near Abernefal. Using map and compass, one has to walk up a very attractive valley in the direction of Gorllwyn. Two things have to be kept in mind. (1) the possibility of a very swift gathering of the clouds and mist; (2) the threat of an inquisitive and seemingly ubiquitous bull. No accommodating trees either.

Not to be recommended to any but the fittest and most stout-hearted — any of these ways — but who wants to be successfully defied by a 'hillock' of only 613m?

Interference of a modern human kind combined with millenia of bleakish weather, has obliterated much of the definite evidence of human activity at the highest point of the Gorllwyn ridge but the remains of two cairns can be made out.

Gorllwyn has a very creditable array of meaningful alignments (see Fig. 10). It has a better mid-summer sunset alignment than has Drygarn Fawr and it has the important mid-winter sunset alignment. It has, what is more, an 090° Equinox alignment which might have made up for the lack of one on Drygarn Fawr and one wonders whether ever at any time these two centres might have been used simultaneously. They are in close and reachable proximity and each could have complemented the other.

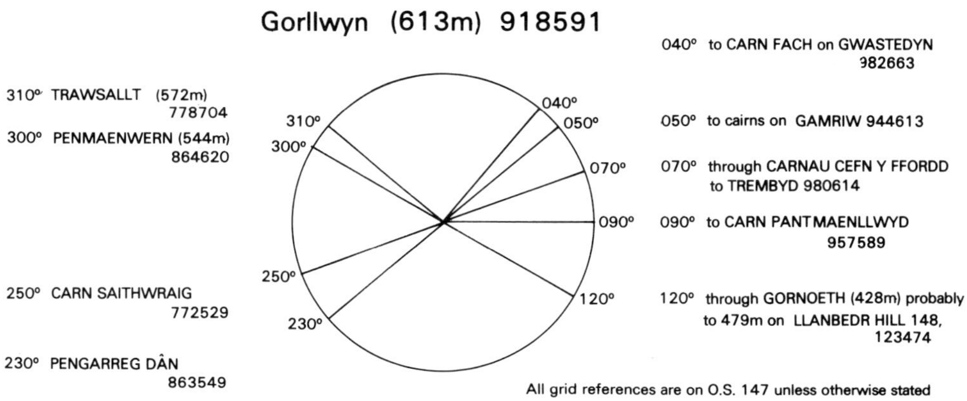

Fig. 10 *The Gorllwyn alignments.*

CHAPTER IX

Gwastedyn

(Fig. 5) (147,986662)

Gwastedyn Hill is privately owned and permission must be obtained from the owner whose farm lies at its foot at Pen-y-ffynnon. Permission is not granted at all times of the year owing to stock problems.

The Gwastedyn Hill site consists of two centres about ¾ mile apart — Carn Wen (477m) and Carn Fach (447m).

Carn Wen shows considerable human activity extending over a lengthy period (Fig. 11). The Carn itself consists of a well-defined circle of stones arranged by human hand with an entrance way in the surrounding earth mound.

Fig. 11. *Stone circle on Gwastedyn Hill near summit of Carn Wen.*

Plate 4: *Circle of Stones on Gwastedyn.*

Nearby there are the remains of a circle within which is a central circle 3′ 6″ or so in diameter formed of flat stones in the manner of paving and reminiscent of those near Drygarn Fawr at 851589 (see Fig. 7b). The traces of several circles marked only by the change in grass colour and vegetation lie nearby. In direct line of sight is Hay Bluff, the Black Mountains, the Brecon Beacons, the Storey Arms gap (in the Beacons), Fan Fawr, Fan Llia and Fan Nedd. Drygarn Fawr is visible as is Gamriw, the Pumlumon and far far in the distance a possible Preseli.

Approaching Carn Fach (447m) from Carn Wen there would seem to be the trace of an old watercourse and an arrangement of big stones extending in a line for a length of perhaps 60 ft. These are not natural outcrop stones and have been purposely placed. As the crest of Carn Fach is reached (lower than that of Carn Wen) there is a northerly view to Pumlumon with the town of Rhaeadr in the deep valley below. Drygarn Fawr is also visible from Carn Fach as is Pont-ar-Elan and Moel Hywel.

Carn Wen itself has been interfered with by the erection of a new and very modern-day cairn built with the best of intentions but probably with the stones of the original of so many thousands of years ago.

On the slopes East from Carn Fach and back along the path to Pen-y-ffynnon are the remains of very ancient human activity in the shape of large circles — broken and scattered but traceable, while modern activity in the shape of bull-dozers would seem to have removed for ever anything else that remained.

Gwastedyn Hill does not seem to have been a solar observatory in itself — but it was certainly on the 070° alignment from Corngafallt and on the 110° alignment from Carn Ricet. These two alignments have been referred to several times previously as being of great importance and particularly revered. Gwastedyn Hill would also seem to have an 060° Moonrise alignment from Drygarn Fawr.

The site possibly played an important role in both Sun and Moon marking alignments.

There would seem to be far more to the story that Gwastedyn Hill has to tell than has been told here and the writer is glad that the site was visited before the bull-dozers absolutely completed their task.

CHAPTER X

Carn Ricet

(see Figs. 5 & 12) (147,870706)

An easy walk for a change

Carn Ricet is on a flat hill-top about 2 acres in area. The Ancient Road marked on the O.S. map of the area runs past it thus making it one of the more easily reached sites. It can be approached from Pont-ar-Elan (904716) and is an easy walk. As mentioned in Chapter II, just before the col leading to Carn Ricet is reached, an ancient track can be seen leading off to the left towards Esgair cywion (881712) and the significance of this has already been evaluated.

There is a cairn on the right-hand side of the Ancient Road on the N.N.E. side of Carn Ricet. From this point, all view southward is

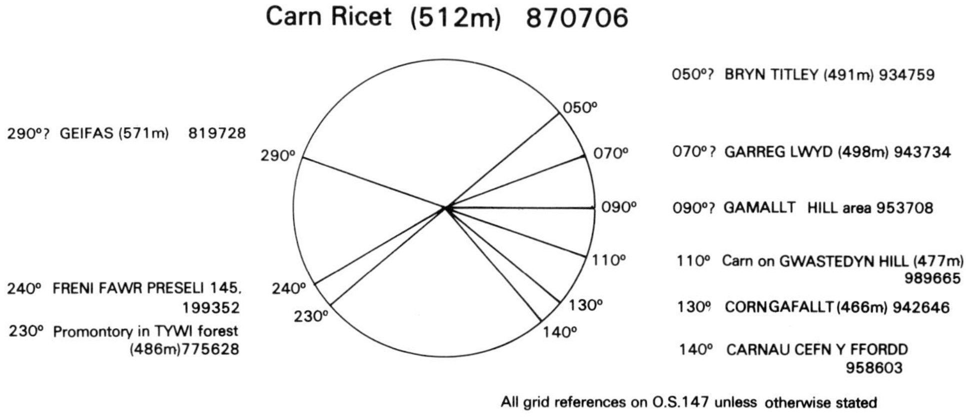

Fig. 12. *The Carn Ricet alignments.*

impossible so its function can only be guessed. It might have indicated the track or given a warning that the site was in close proximity. Here quartz boulders are to be seen in the stone arrangement of the cairn. Owing to the accessibility of the site, there is little left to see — but much might perhaps still be hidden under the thick peat layer that envelops the area.

Freni Fawr in Preseli in the far distance is seen from Carn Ricet just as it is seen from Drygarn Fawr and provides an alignment of 240° (see Fig. 12). The use of Carn Ricet has been substantiated by certain meaningful alignments but work on many of its outlying sites has been held up mainly because of inaccessibility. Investigation is however continuing to be carried out and it is hoped before long to have reached all alignment markers.

In addition, studies are being made at present on a number of long barrows (possibly Neolithic) that the writer has discovered in this area.

CHAPTER XI

The Carneddau

(see Fig. 5) (061532 area)

Geological Background

The Carneddau range of hills form the southern end of the Builth-Llandrindod Inlier of Ordovician rocks. In this region, muddy deposition during Lower Llanvirn times was followed by intense volcanic activity when some 1500 to 2000 feet of spilitic and keratophyric lavas and ashes were erupted. It is this thick Builth Volcanic Series which is responsible for the rugged high ground of the Carneddau range. Caer Fawr is formed of spilites and keratophyre whilst the higher summits (marked 436 and 444 metres) to the north-east of Newmead Farm are formed of bouldery spilitic ashes. A number of dolerite instrusions also occur, as for example on Caer Fawr and Cae Einon. Particularly interesting are the deposits of felspar sands and boulder beds (the Newmead Series) which are believed to represent the actual coastal deposition around an emergent volcanic area during Upper Llanvirn times.

The region is intensely faulted with several E-W fractures breaking the continuity of the mainly westward-tilted volcanic horizons. Other important faults trend NNE-SSW along the range.

(I am indebted to Professor T. R. Owen for this brief geological background description of the Carneddau.)

Promontory (392m) Caer Fawr (058531) and promontory (436m) (063546) above Newmead Farm seem to have been the two centres from which meaningful astronomic alignments were available. These sites are easy of access but permission must be obtained both from Newmead Farm and Cilberllan to cross this private land and great care must be taken to observe the Country Code in every way — especially in regard to dogs as this is sheep-land.

From (436m) 063546 there is a line to Hay Bluff (161,244366) of 140° and a midsummer sun-set to Corngafallt. There is too an important mid-winter sunset directly along the ridge of Eppynt to Trichrigiau and on through a line of cairns (see Fig. 5). This area cannot be visited as it is in an Army danger area — which is a pity — but each era in the life of Man has its attendant necessity. The fore-knowledge of a mid-winter sunset to the one, an artillery-range to the other — in the same location but eight thousand years or so later.

[In the course of fieldwork that is being undertaken in the area of Herefordshire and the Black Mountains, it has been noted that there is a midsummer sunrise reading from a point on Hay Bluff to the hill near Dorstone on which is situated the megalithic tomb known as Arthur's Stone. This is a chambered cairn considered as dating to between 4000 and 5000 B.C. There is a Standing Stone nearby which could itself predate the tomb and also be part of the astronomic alignment to Hay Bluff.]

From (392m) Caer Fawr there is an important midsummer sunset to Drum-ddu (969603) at Carn y Geifr. Caer Fawr is very interesting to visit as a fortification of much later times but there would seem to be nothing left of what doubtless had been there in the times under consideration.

CHAPTER XII

The dawning of rites and sacrifice

"Le plus ça change . . ."

As the reader is aware, the period under discussion spans a time-length of about four to five thousand years. This is already twice as long as the period that separates us from the date of the birth of Jesus Christ. This fact does make us able to realise that those peoples who were in existence at the beginning of the period would differ in many ways from the peoples of one or two thousand years later and even more from those peoples of two or three thousand years later still.

We have already reached a point in this book where we are talking about people who existed some few thousand years later than those about whom we were talking in the beginning.

We now have small communities, primitive just the same, pagans but not savages, and comparatively better organised in that they even have taken the foretelling of the approaching seasons as accepted certainties around which they planned their lives. Catastrophic forest fires started by lightning or by human carelessness would still work havoc, as would flooding, disease and other natural calamities. Bad Spring weather blew blossom from trees and the Autumn fruit and nut harvest would suffer as a consequence. It seemed as if some form of appeasement of the superior powers who governed such events might secure more favourable treatment and conditions.

So many elements seemed to be required to be appeased — the Sun, the trees, the water, the wind. Where were these to be appeased but at the sites of the observatories?

It is easy to appreciate how what began as a simple attempt at the understanding of the comings and goings of the Sun to help daily life, while developing in time into an astronomy of quite some sophistication developed, too, into a Sun worship with all the pagan rigmarole that that entailed.

It is only a short step before those who imparted the knowledge gained from the time calculators (or Sun-observatories as they are also called) were in the position of men of power. They were in a position to instil reverence, to threaten and to create fear. They were the new social caste of astronomer-priest. The more sophisticated the astronomy, the more sacrificial the Sun-worship would tend to become. It is impossibe to divorce the two trends — the one gave rise to the other and they are inextricable.

It is quite clear that only the "brighter boys" would concern themselves with the operations of the Solar observatories and they took care to keep the information in "restricted circulation" — the role of astronomer-priest even becoming restricted to a hereditary family relationship.

As sacrifice and offerings are the visible forms of appeasement, so is natural fear, bred from ignorance, the catalyst. The astronomer-priests were in a position to exploit it. And this they certainly did. Knowing full well that the Sun would rise from behind a certain hill on a certain day, they were able to pretend that if suitable offerings were not made, then the Sun would not appear. Such a possibility would terrify their small audience who would be well aware of the consequences of such a calamity. Offerings would immediately be made and the appearance of the Sun would be awaited, not without some anxiety. And, appear it did — as it would have anyway, but wonder, faith and belief had been created and here was a religion.

It is a little discouraging when reflecting on the human state, that as much as perhaps five thousand years later, the Romans, so respected for their advanced civilisation were doing exactly the same thing! Vergilius Maro in his Book III of the Aeneid makes Aeneas say (to nobody's surprise at all) —

> *"Sacra Dioneae matri divisque ferebam*
> *auspicibus coeptorum operum superoque nitentem*
> *caelicolum regi mactabam in litore taurum."*

"I was sacrificing to my mother Dionae and to the gods to be auspicious to the works I had begun and to the heavenly king of the dwellers in the sky, I was giving a fat bull as sacrifice on the shore."

"Le plus ça change, plus c'est la même chose."

CHAPTER XIII

A Mesolithic Depopulation of the Uplands and the Path to Preseli

(see Fig. 13)

As what were originally such simple seasonal clocks developed into observatories of a greater degree of sophistication, the astronomer-priests in charge of them began to nurse ambitions of entering upon more highly advanced scientific exercises. The Moon had taken their interest and the Moon did in one month what the Sun took a year to do. The Moon had eclipses. Here was a challenge to their skill in being able perhaps to foretell these eclipses and would it not incidentally increase their power over their gullible and cowardly flock if they could use their pre-knowledge of an eclipse to instil even more terror and even greater obedience to their demands?

The terrain around Drygarn Fawr and the other observatories in the area did not lend itself to any such experiments and while, having been pre-eminently suitable for the requirements of the past owing to the proliferation of hill-tops furnishing the requisite alignments, these very hilltops and their haphazard topographical arrangement put the astronomer-priests at a disadvantage in the exercise of their more sophisticated requirements.

Word had, for some time, been getting through of a possibly more successful observatory situation behind the Freni Fawr tump which had been lying so curiously and so far down in direct line of sight beyond the Bryn Brawd cairns.

From this direction came the very ancient early February and early November sunsets, alignments important in their very antiquity, from which developed the Celtic festivals of Imbolc and Lughnasa, ancient in themselves — from which in turn comes the Celtic New Year, still celebrated in the Gwaun Valley in Pembrokeshire. The early November sunset, the Calan Gaeaf (the first of Winter) is itself still widely celebrated with cakes and candles placed inside scooped-out swedes — the traces no doubt of the old rites and offerings mentioned in a previous chapter.

The area intervening between that of Drygarn Fawr and the Preselis would undoubtedly have been exploited in the hunting season by Mesolithic hunters from both ends over several millenia, for there had been Maglemosian inhabitants in the Preseli area before the slopes of Drygarn Fawr had become sufficiently temperate to permit of human habitation (see Chapter I). Then, when a milder climate in the upland regions had established itself, families of Mesolithic hunters had occupied the upland areas living thus contemporaneously with those in the Preseli area.

Pursuing a similar way of life dictated by similar conditions of habitat, they were in all probability for long spasmodic periods unaware of the existence of each other. As the years passed, parts of the connecting area would have had settlements rather longer than seasonal in their nature and while the population showed no great increase from the times of 7,500 B.C. when it was about 10,000 in the whole of the British Isles, intermittent contact was probably made between hunters from the two ends of the ridge as these settlements allowed the hunters to increase the scope of their seasonal hunting grounds.

The hunters from the Preseli end would have seen the massive Drygarn Fawr from behind which rose the early May and early August sun (Dates of importance reaching back, as Dr Margaret Murray says, to Upper Palaeolithic times occurring as they do between the solstices and the equinoxes). But a considerable length of time would pass — probably in the region of four thousand years — before anything at all resembling what may be called a *route* from one area to the other was in regular use.

The later Mesolithic peoples had realised for some time that a more settled form of life was now possible and was more acceptable. A certain degree of knowledge of what might be termed animal husbandry had been acquired and it had been learnt that meat for food was readily obtainable without having to run after it. A primitive sowing of seed and a reaping of the results had also been learnt and this new knowledge led to a change of attitudes and to a gradual change in the way of life.

If signs there are in these Welsh Upland areas around Drygarn Fawr of any form of settled agricultural life at this time, then they are remarkably few, for the lands lower down and farther away to the West, the South and the East provided less harsh conditions to this end. The heights, among them Corngafallt, with its human seasonal occupation by hunters over so many millenia were being gradually abandoned in favour of establishing hill-top settlements of a farming nature on lands at a lower level. This is the way of life that, developing gradually from the Mesolithic unsettled hunting way of life, helps to distinguish it from what has been accepted as the Neolithic way of life.

The Path to Preseli (Fig. 13) is easily distinguishable even today and Modern Man still puts parts of it to his own use, as, for example, it runs for some miles over Mynydd Llanybydder past Crugiau Edryd and through Crug y Biswal — and again, sometimes as a lane, sometimes a track, then disappearing and reappearing as a roadway all the way from the general area of Bryn Brawd cairns to Mynydd Pencarreg.

It must be appreciated in examining Fig. 13 that many of the carns named in the sketch of the route were not in existence when the earliest travellers trod the path and that many belong to later and successive waves of settlers. They have been recorded here merely to make it easier to recognise the general route of the Ancient Road.

As a path, in the broadest sense, it came into existence originally when the need for it came into existence. This was at the time when the communities of hunters in the upland areas, finding it impossible to adopt a more settled way of life in the harsher hill regions (partially through their own fault — in deliberate forest firing for clearance purposes and other forms of despoliation through ignorance) drifted to more acceptably mild areas such as Pembrokeshire. It was along this ridge in a direct visual line to their goal that they made their way. Travelling along a ridge was still the wisest course, for to travel along higher ground provided greater security and, weather permitting, enabled them to keep their destination in sight at all times.

Though forest fires would have cleared great tracts, deciduous

forests still covered the ridge which is nowadays naked, except for plantations of conifer planted by Modern Man. Wolves, boars and other predators were still a peril to the unwary traveller so that travelling the Path at any time was still fraught with danger.

The idea of their travelling the Path in one long continuous journey is, in view of the times, incongruous. There had to be days of rest, there had to be days to hunt and collect food. Some would forget about wanting to go all the way and would establish a community somewhere en route. Some would die. Some would eventually make it to the end.

Some of those who did make it to Preseli were, of course, the astonomer-priests who being frustrated with the constraints of local topography had determined to set off to the promise of the extension of the parameters of their astronomic expertise.

Neither the astronomer-priests nor the late Mesolithic and fringe Neolithic groups emigrating westwards from the Drygarn Fawr area in the approximate times of 3,500 B.C. to a more settled way of life "farming in Pembrokeshire" were likely to have met on the path any of their contemporaries from Preseli engaged in a similar operation in the opposite direction That the Path formed part of the route that travellers ultimately took to go East (and to Salisbury Plain, who knows?) is certainly a great possibility, and the Ancient road that runs along the entire length of the Preseli ridge was in all probability linked by this very path from Freni Fawr in the direction of Carn Ricet to the Ancient Road shown on O.S. 147 winding its way brokenly eastward through the area of Pont-ar-Elan.

Thus, just as some of their earlier Mesolithic forebears (see Chapter I) so many thousands of years earlier had, as simple explorative hunters slowly travelled the ridge up from Preseli towards Drygarn Fawr on the direct line of sight towards that so significant May sunrise, so now some of their later Mesolithic descendants travelled the ridge back to Preseli. Not as the simple explorative hunters of an early post-glacial period did they now journey back but, in the main, as astronomers of some experience, some thousands of years later, nursing hopes of advancing their scientific astronomic expertise in an observatory already long established and based on what they thought to be a less frustrating site.

CHAPTER XIV

Preseli – The Unfinished Symphony

(Figs. 14, 15 & 16)

Prelude to Stonehenge
The Opening Bars and Variations on the Theme

Geological Background

The rocks of the Preseli Hills are of Ordovician age and fall into three main categories. The Ordovician sedimentary succession comprises 3500 feet of blue, grey and black slates, shales and mudstones. Some grits and quartzites appear near the top of the sequence, mainly on Freni-fawr. Volcanic rocks are developed at two horizons within the shales and slates and comprise up to 700 feet of spherulitic rhyolites, trachytes and volcanic tuffs. The rhyolites form Foel Drygarn and Carnalw. Thirdly, the sedimentary and volcanic rocks are intruded by a number of thick quartz-dolerite sills, which trend E-W across the main range and form marked serrated crags such as Carnmenyn, Carnbica and Carnbreseb. These dolerites are often characterised by the presence of numerous pink and white spots (mainly of felspar) up to 2 cms in diameter (the "spotted dolerites" of H. H. Thomas and almost certainly the source of the Stonehenge "bluestones"). (The spherulitic rhyolites, too, can be matched at Stonehenge.) All the rocks of the Preseli Hills have been folded along WSW-ENE axes by Lower Palaeozoic earth-movements. A number of eastward-pitching anticlines and synclines occur, accompanied by southward-directed overthrusts.

(I am indebted to Professor T. R. Owen for this summary of the geology of the Preseli Hills.)

Plate 5: *The Nant-y-Groes axe-head showing its size relative to a £1 coin (see page 89).*

Having left, in the last chapter, our peoples of the Welsh Uplands making their way slowly through the centuries along the ridge back in the direction of Preseli, it would be as well in the beginning of this chapter to go back and remind ourselves of the conditions prevailing in Preseli from the end of the Ice Age.

While in the last (Devensian) glacial episode of the Pleistocene Ice Age, the Preselis themselves would have been buried under ice sheets, there were, nevertheless, in quite close proximity to them and, especially southwards, large areas of what was frozen tundra. When the general amelioration in the climate which brought about the slow end of the Ice Age occurred, these areas and the Preselis themselves would have cleared of ice earlier than the higher more mountainous areas inland like the Drygarn Fawr area could be presumed to have done.

Early Mesolithic Man would therefore have been existing in these areas for quite some time before he made his appearance in the Central Wales area. As Brian John (1984) in his interesting book "*Pembrokeshire*" says, finds of Mesolithic tools have been made at Nab's Head (790110), Benton (on the Cleddau) and on Frainslake Sands (889980) west of Castlemartin.

Owing to the fact that the Devensian Ice sheet left areas of South Pembrokeshire free of ice, an uncomfortable form of human existence was possible and in the southern part of Pembrokeshire, as Brian John points out, evidence has been found of the habitation of caves by small family units of Mesolithic people in Pembroke as well as in the Tenby-Caldey area at a very early time in the post-glacial era — possibly being as early as 9,000 B.C.

As time went on and a greater amelioration took place in the general climate, later generations of these peoples ventured farther abroad in the area and would have spread their hunting activities up into the area of the by now tree-clad slopes of the Preseli ridge.

It would not be unreasonable to assume that the needs of these Mesolithic people to be able to foretell the seasons would have been met by the establishment of simple seasonal clocks on such promontories on the Preseli ridge that were able to perform this function. Foel Cwm Cerwyn (536m) being most outstandingly the highest was probably the earliest.

Being as simple as we have seen the Drygarn Fawr seasonal clock to have been in its original form, the observatories on these promontories in Preseli were, in the course of the millenia, to benefit from all that had been learnt in the observatories of Central Wales and were to become the last grounds for experiment before the establishment of the final magnum opus on Salisbury Plain.

Only a brief glance at the remains of the Preseli observatory shows that, in its hey-day, it was more complicated and sophisticated than any of the observatories of Central Wales ever were. (See Figs. 14 & 16.) It probably started somewhat earlier than they, it existed contemporaneously, and it was still in operation when they had passed their peak and were being gradually abandoned. Benefitting from knowledge gained in these centres, it represented an advance on them and it is a matter for great regret that owing to such a high level of agricultural activity in the area, from Neolithic right up to Modern times, much has undoubtedly been erased.

Note: The attention of the reader is drawn to the need not to confuse Foel Drygarn (158336) in Preseli with Drygarn Fawr (863584) in the Welsh Uplands.

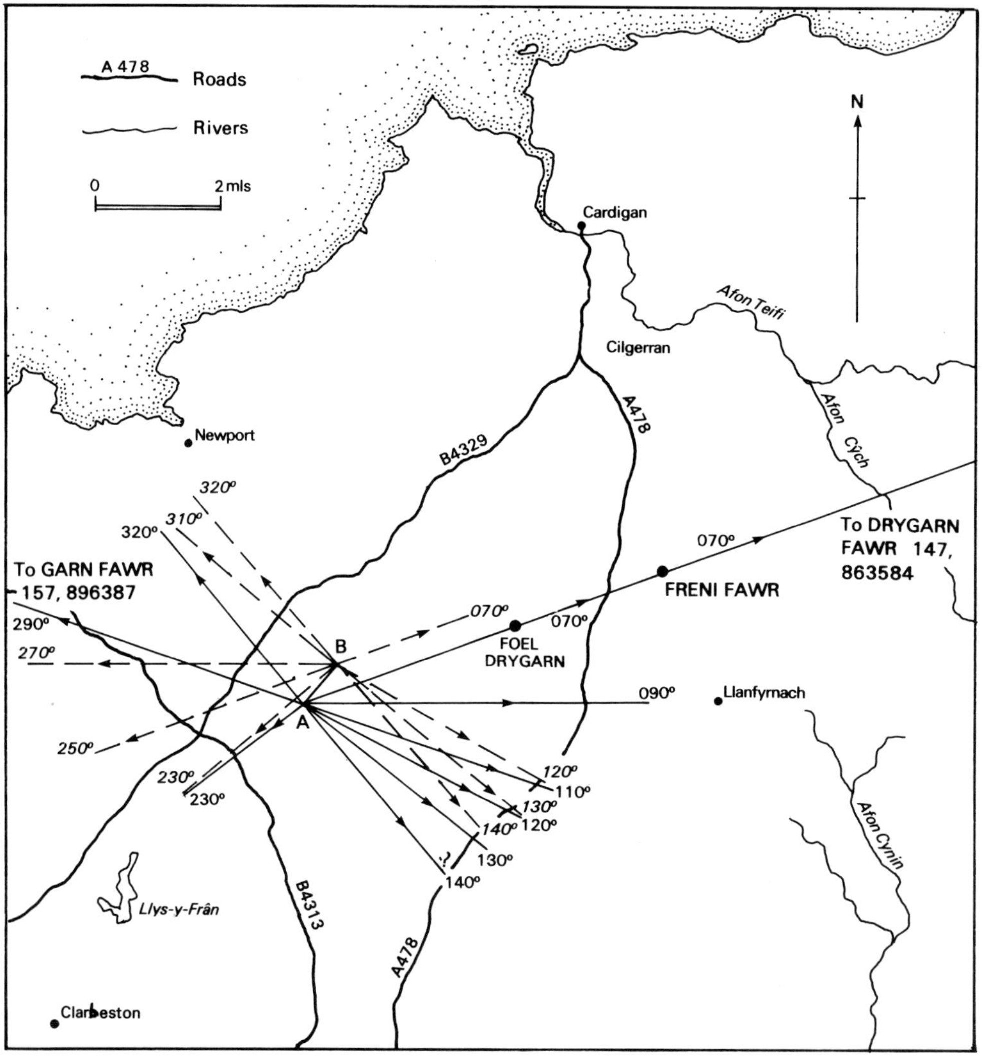

Fig. 14. *The Preseli Complex. This does not represent all the complex features of Preseli – and should be studied in conjunction with Figure 16.*

I. *Foel Cwm Cerwyn (536m) (Fig. 15)* (145,094311)

The presence of cairns, and ditch and mound circles on the summit indicate human activity.

1. Considering the alignments from Foel Cwm Cerwyn, there is an 040° alignment straight to Foel Feddau with the latter furnishing therefore a reciprocal of 220°.

The study of the Moon began later than did the study of the Sun which body had a more obvious seasonal significance. The setting up of a lunar calendar which would keep in step with the seasons was far more difficult, too, as the twelve months of the Moon are shorter than one year by 11¼ days. When this error was allowed to accumulate, Caesar, in reforming the calendar, had to have one year of 445 days to correct the imbalance. Since that time, while retaining twelve divisions of the year, we have had to make them longer than the lunar month by introducing some months which have thirty-one days.

Such then were the problems facing any of the astronomer-priests of later Mesolithic times who had ambitions to set up a lunar calendar.

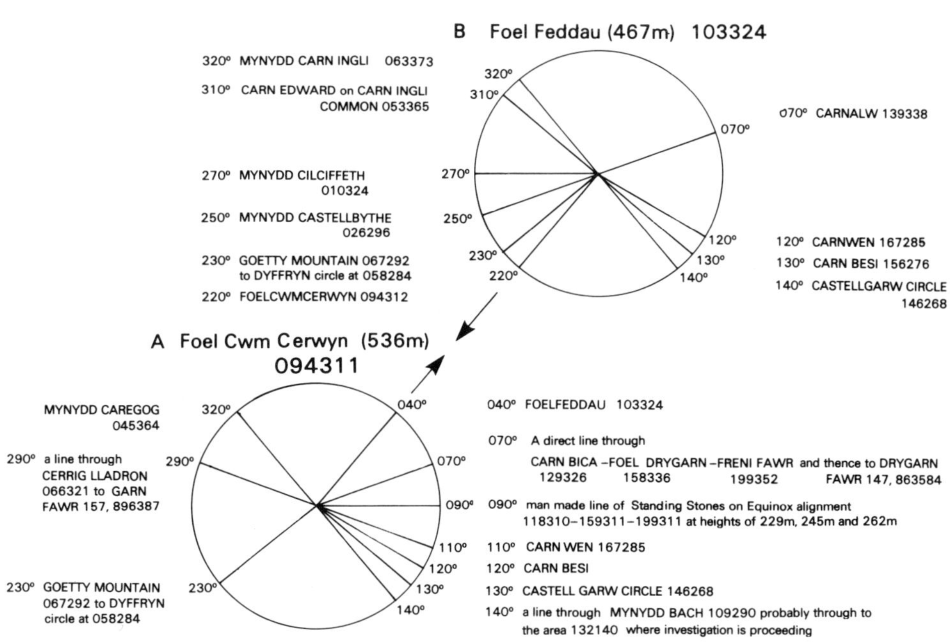

Fig. 15. *The Preseli alignments.*

Having discovered where on the horizon the Sun appeared to rise and set, their investigation of the positions of the Moon's rising and setting might originally have been nothing more than the pleasure of an intellectual exercise. It is in the nature of Man to wish to improve on his earlier accomplishments. Perhaps it was that the astronomer-priests, getting bored with one success were playing around. Or perhaps were their playings around nothing to do with the establishment of a lunar calendar at all and their aim lay in some other direction.

2. The 070° Sun alignment through Carnbica to Foel Drygarn and thence to Freni Fawr is one of great interest. Not only owing to the attention that Dr Margaret Murray drew to the importance of such an alignment as 070° from Upper Palaeolithic times but also to the fact that the early people of Preseli judged it important enough to build three large cairns on the summit of Foel Drygarn on an 070° alignment of a sun-rise which pointed in the direction of Freni Fawr and thence up to Drygarn Fawr itself. It is significant that there are signs of human habitation on the S.E. slopes of Foel Drygarn in the form of three just discernible circles marked now by changes in vegetation. This was an important time of the year — early May, a time for rejoicing at the coming of summer and not a solstice or an equinox so therefore dating from most ancient times and one from which the great Celtic festival of Beltane sprang as did the custom of dancing around the maypole that takes place today.

3. The 090° Equinox line (see Fig. 16) is of more than usual interest as we have here a line established by the hand of Man himself. Getting by now more and more weary, no doubt, of having to depend on nature's topographical whims to provide suitable or para-suitable astronomic alignments, three separate sets of Standing Stones were erected on this line with a remarkable degree of accuracy even to today's sophisticated standards — as the grid references themselves show — the northings reading 310, 311, 311 from Foel Cwm Cerwyn which has a 311 northing itself. Two other points of interest to note are:

(a) This is the first example we have found so far in the examination of the observatories, of Man himself making and marking an alignment with stones in the absence of a natural topographical alignment. That he should have marked a suitably aligning promontory by a pile of

stones, which is all we have found he did so far, is a different matter altogether. By this action, he is showing us that constraints of Nature are not going to be allowed any longer to deprive him of alignments that he considers necessary (see end of this section).
(b) The Standing Stone at 207311 in the area of Bigni, (262m) (199311) is at an approximate distance of 2½ miles from those at 159311, and the Standing Stones at 159311 are the same distance approximately (2½ miles) from 118310. They are therefore at an approximately equally spaced distance from each other. This shows us that Man could now measure quite a considerable distance with a fair amount of accuracy without the markers being even within sight of each other! As the Preseli area has been subjected to much human disturbance, mainly agricultural, over a far longer period right up to modern times than any such comparable activity in the Drygarn Fawr area, there is a great sense that much has been lost that could tell an even more interesting story than the one above — perhaps solving the enigma that will forever cloud the full understanding of what the function of Preseli was in its finality before being partially dismantled and carted off to Stonehenge in a form of Neolithic D.I.Y. kit. (See Standing Stones, a closer evaluation, page 82.)
4. The 110° alignment, an early February and an early November sunrise crosses a 120° moonrise minor southerly standstill from Foel Feddau at Carn Wen, as in a similar way the 120° moon minor southerly standstill from Foel Cwm Cerwyn crosses a 130° midwinter sunrise from Foel Feddau at Carn Besi, just as a 130° midwinter sunrise from Foel Cwm Cerwyn crosses a 140° moonrise major southerly standstill from Foel Feddau at the Castell Garw Circle to Llanmadoc Hill in Gower (159,430923 area).

That there would be crossings of alignments from closely neighbouring centres of observation would in the very nature of things be inevitable but what is curious is that there is nothing to indicate anything deemed worthy of marking farther along the alignments in the surrounding countryside — as if it was known that that is where they coincided and that is where the markers were placed! Was the marker placed at the crossing of the Sun and Moon alignments with any intention or was it fortuitous? There is a 140° moonrise major southerly standstill alignment through Mynydd Bach but a more satisfactory marker farther along the alignment does not exist in the present day and might have been lost.

5. There is an important 230° midwinter sunset marker to Goetty Mountain. This was an important point in the solar calendar as it indicated that the shortest day had been reached and that from henceforth the hours of darkness would decrease in number. This alignment coincides in prolongation with the 230° alignment from Foel Feddau which again would be a geographical inevitability in view of the positions of the relative centres.

6. The 290° May and August sunset is again one of the important ones that disregard the solstices and the equinoxes and belongs to those that date from the most ancient of times. The sunsets were of course more important than the sunrises in this class of alignment, as night to those peoples was more important than day, and day was just something that followed night — as has been said before. The 290° alignment from Foel Cwm Cerwyn passes through Cerrig Lladron to Garn Fawr (157,896387) and that sunset would have been spectacular in clear weather with the beyond-lying-sea lending greater intensity to the Sun's dying rays.

7. The 320° alignment to Mynydd Caregog represents that of a moonset, a major northerly standstill. Mynydd Caregog is a 311m rise on the slopes of Carn Ingli Common — an area showing the traces of considerable human activity belonging to this period.

The Standing Stones at 090° from Foel Cwm Cerwyn – a closer evaluation (Fig. 16)

Much work as been done in the Preseli area that cannot be included in this edition largely because it is intended to treat certain features of the Preseli observatory only in their relevance as a stepping-stone between the simple Drygarn Fawr and what were to be the far more mathematically calculated positions of Stonehenge.

When examining the accompanying diagram (Fig. 16), thought must be taken, as it is only too easy to confuse constructions from different periods. While noting the position of churches for example and the siting of Tumuli and Standing Stones alignments, it must be remembered that thousands of years separate these events in spite of the fact that a certain relevance exists between the siting of a

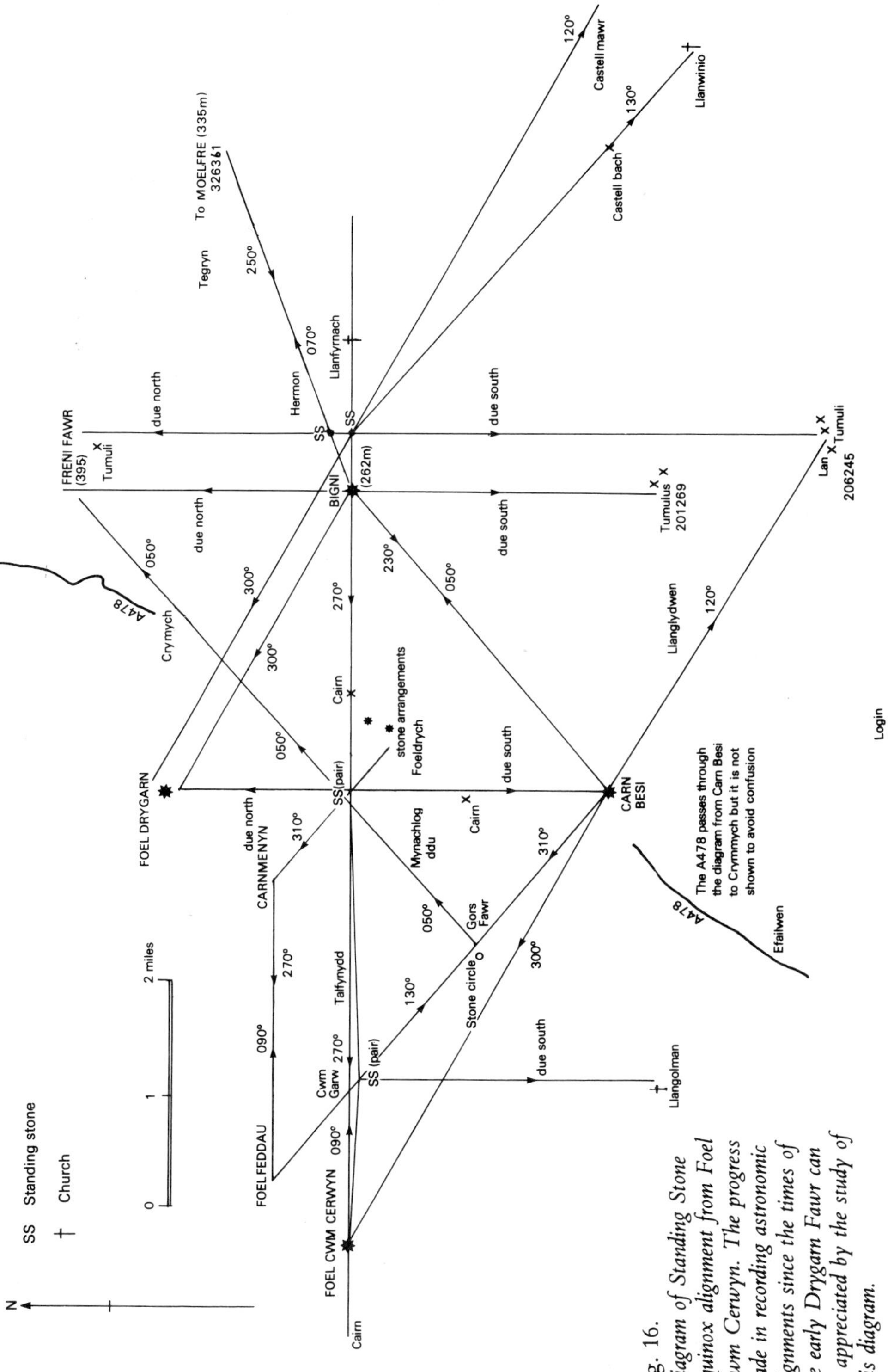

Fig. 16.
Diagram of Standing Stone Equinox alignment from Foel Cwm Cerwyn. The progress made in recording astronomic alignments since the times of the early Drygarn Fawr can be appreciated by the study of this diagram.

Christian Church and the place of possible pagan worship in a bygone age. It is known that there are many examples of the building of a Christian place of worship on the site of a previous pagan place of worship — a few examples of which are shown on the accompanying diagram (Fig. 16).

The Standing Stones in the areas of 118310, 159311 and the area of 199311 would seem to serve as directional wayfaring marks from Foel Cwm Cerwyn in an easterly direction all the way to Bigni — (and of course, reciprocally westward from Bigni to Foel Cwm Cerwyn). Those on Bigni at 207311 serve as directional markers in a due northerly direction to Freni Fawr, since they are no longer on an alignment proceeding farther east but on an alignment to each other pointing north to Freni Fawr. Thus we find Man now making use of naturally occurring topographical features (as in the Drygarn Fawr area) and combining them with stones erected at specifically calculated points on a meaningful astronomic alignment. This is far in advance of anything we have seen in our studies in the Drygarn Fawr area while not quite yet attaining the stylishness of a Stonehenge.

Examining first the Standing Stones in the area of 118310, they are found to be slender, of spotted dolerite (bluestone) about 7 ft. tall and at the foot of Foel Cwm Cerwyn.

(1) If, on leaving them and walking in an easterly direction, a sighting is kept on their alignment so that they appear to be a *single* Standing Stone, then the true due easterly path is being followed. As soon as any deviation from the true easterly bearing occurs, the stones become out of alignment to each other and so *two* stones are seen. In what better way could these peoples of thousands of years before Christ still be speaking to us?

(2) These Standing Stones are on a due south alignment to Llangolman where a Church is sited — the significance of this is worth consideration, while bearing in mind at the same time, the span of time referred to above.

(3) The stones are also on a 130° alignment from Foel Feddau to Carn Besi (reciprocal 310°).

Examining the next pair of Standing Stones in an easterly direction — those in the area of 159311, it is found that (1) they too are of spotted dolerite (bluestone) and while in themselves furnishing an East-West alignment are in the path of a North-South alignment from Foel Drygarn to Carn Besi *over* Foel Drych which is plentifully furnished with broken, obviously man-constructed stone arrangements (rhyolitic ash) lying long hidden in the heather providing the appropriate directional information. The top of Foel Drygarn is visible just over the tops of the trees growing on the intervening rise. (2) An 050° alignment from the stone circle at Gors Fawr 134293 to Freni Fawr, have these Standing Stones in its approximate path at the point where they have been sited to mark the East-West alignment. If they had been sited anywhere else on the East-West alignment, they would not have been positioned on the North-South line from Foel Drygarn to Carn Besi, the 050° alignment from Gors Fawr to Freni Fawr and the 310° alignment to Carn Menyn from a point on Foel Drych. Their positioning, therefore, where *eight* alignments coincided must have been most carefully calculated and shows some considerable expertise.

(cf. the position of SS at 118310 being on five coinciding alignments and the position of SS at 207311 being on eight coinciding alignments.)

In examining Bigni (199311) 262m.
(1) The information immediately above explains why the SS (Standing Stone) of spotted dolerite on the western slopes of Bigni is not on the top of Bigni but at 207311 at 198m. It had obviously been realised that if it had been on the top of Bigni, certain requisite alignments would not coincide at that point. This realisation in itself indicates great progress in the field of astronomic alignment even if originally there had been something of a hit or miss situation. Here, Man, finding that the topography did not exactly answer his needs, substituted Standing Stones in the precise position relative to his requirements. One can imagine the talk that went on among the builders of the time when it was found that Bigni was not, after all, as suitable for their requirements as first thought. On the other hand, Bigni might very well have been used by earlier builders for a 230° alignment to Carn Besi with an 050° reciprocal. Remembering what was written above, constructions from different periods could cause confusion in interpretation of what we see here! The fact, however, remains that

Plate 6: *The Standing Stone at 207311 with the writer standing on the path of its North alignment to the Standing Stone at 207314 and thence to a point on Freni Fawr, both the latter being obscured by the tree and other vegetation.*

Plate 7: *The Standing Stone at 159311 which has a 310° alignment passing through it from a point on Foel Drych to Carn Menyn on the horizon behind which the sun marking the longest day would sink.*

(a) Bigni is on a N-S alignment — from a point on Freni Fawr to Tumuli at 201269. (b) The SS at 207314 and 207311 are similarly on a N-S alignment — from a point on Freni Fawr to Tumuli at Lan on 206245. This tells us that possibly at some time thousands of years ago the one site or the other failed to give complete satisfaction. At a guess, this could have been Bigni — and the two parallel N-S alignments in such close proximity to each other, as is seen on the diagram, could well be a reflection of consternation, conversations and arguments that took place in early Neolithic times!

(2) Bigni itself on the East-West alignment beside being on the North-South alignment mentioned, has a 230° alignment to Carn Besi from itself (thus providing from Carn Besi an important 050° reciprocal).

(3) There is also a 300° alignment to Foel Drygarn. The 230° alignment mentioned in (2) and this 300° alignment could indicate that possibly at one time there were several others that, if they were not to topographically aligning points, have been destroyed.

In the examination of the Standing Stones at 207311 on the land of Mr & Mrs A. D. Chaplin of Nant-y-Groes, (1) it is found that it is of spotted dolerite — as are all these Standing Stones in this study. It is very tall — about 9 ft — and stands astride an East-West alignment and was once one of a pair. The easterly alignment continues to a tumulus at Llanfyrnach in the almost immediate proximity of which is a church. It would seem that if any farther alignment to the east had ever existed, no trace by now remains. (2) It is also astride a N-S alignment from tumuli on Freni Fawr to tumuli at Lan (church, in translation) at 206245. (3) There is a 300° alignment to a point on Foel Drygarn and there is a 120° alignment to earth-works at Castell Mawr at 271277 and a 130° through Castell Bach to Pant-y-Maen (the slope of the Stone, in translation) at 266260, which line passes through the site of a church at Llanwinio 261264. It would not be unreasonable to come to the conclusion that in the area under immediate study, we have examples of the siting of four Christian places of worship which in all probability overlay a pagan site of thousands of years earlier.

In examining the SS at 207314 on the land belonging to Mr and Mrs E. N. Jones of Bryn Meini, this stone is a full 12 feet tall, most impressive and of spotted dolerite (bluestone). This is one of a pair —

its partner having been removed about 40 years ago. As this is within living memory, Mr Jones was able to show me where its partner had been positioned. This was not on an E-W alignment but on a N-S axis. (If the alignment of the missing partner of stone 207311 was a N-S or E-W one, will never be known as there is no trace of its position — although what is believed by Mr Chaplin to be its shattered remains lie piled in a corner of a field on his land.) It has also to be noted that the SS at 207314 stands on a direct 070° alignment from Bigni to Moelfre (335m) 326361.

It requires only a very casual glance at Fig. 16 to realise the great progress that Man's ability to calculate astronomic alignments had made even from the simple but highly satisfactory compass-style working of Foel Cwm Cerwyn and Foel Feddau and certainly from the similar style-of-working Drygarn Fawr. In all these cases topographical features were used for aligning the sight to the distant horizon.

As his needs spread more into the intellectual field, the degree of sophistication required becoming correspondingly greater, we find Man marking alignments by arranging stones in easily visible places where nature had not sufficiently appropriately met his requirements. One of the first examples of this new trend can be found in the stone arrangements on the long ridge of Gamriw — one arrangement at 147,944613 marking a midsummer sunrise from Gorllwyn (see Fig. 10) and another arrangement of two piles close together at 943614 marking a 180° from Corngafallt (see Fig. 8). The recording of this latter reading being, in itself, an indication that it was the work of later Man, as to earliest Man it was not one of importance. This then, was indeed the beginning, the beginning of the move away from the topographically aligning hill-top, with its inherent constraints, to the freely placed and well-calculated man-positioned marker.

In his circa 2,500 B.C. phase, we see Man, much later and more experienced now, erecting Standing Stones (1) to mark astronomic alignments where topographical features proved inadequate and (2) combining them with topographical features of astronomic significance on a large and complicated scale — innumerable examples of this being found in Fig. 16.

Bigni (262m) would seem to have already been playing a part of some astronomic significance for sometime probably before Man, with ever increasing expertise, erected his Standing Stones. No longer now were they to be erected as wayfaring signposts pure and simple, but they were to be calculatedly positioned at the virtual exact crossing points of not just one but of several astronomic alignments known to exist between one topographical feature and another.

That this should happen in the case of each set of Standing Stones shown in Fig. 16 can be no accident — especially as the Standing Stones, all having an E-W alignment in common, could very well have been erected absolutely anywhere along this 090° alignment from Foel Cwm Cerwyn. The fact is, they were *not,* and the sole conclusion that can be drawn is that the exact spot where they *were* erected was pre-calculated, their positioning deliberate.

This conclusion is one in which Modern Man should find great delight and one in which he can take great pride, for the Man who executed this complicated and precision-demanding exercise so many thousands of years ago is the one and self-same who today can land on the Moon and fly a space-shuttle.

(As a result of an interesting and informative conversation with Mr and Mrs Chaplin of Nant-y-Groes Farm, on which land SS 207311 stands, it appears that there are a considerable number of man-placed stone arrangements on the Nant-y-Groes land. In one field alone, there were the remains, very apparent to the writer, of five separate circles of unusually large stones and in another field, in which remains of stone arrangements had been piled (not by Mr Chaplin) unceremoniously along a hedge, an interesting pattern was found by the writer on one long slender stone (see Fig. 17). In relation to this, it was remarked that some of the unusually large stones in the circles mentioned above also bore man-made markings in the form of grooves of varying direction. A photographic record was made of these.

Mr Chaplin showed the writer a truly magnificent example of a flint-core axe — about 7″ x 3½″ (see plate 5) which he unearthed in 1984 from not more than 2 ft below the surface of one of his fields. The edges still have a sharp bevel and it gives the appearance of having been made only yesterday. This implement compares favourably with those of later Upper Palaeolithic Age (Aurignacian — Proto-Solutrean) thus giving it an approximate age of between 18,000 and 35,000 B.P.)

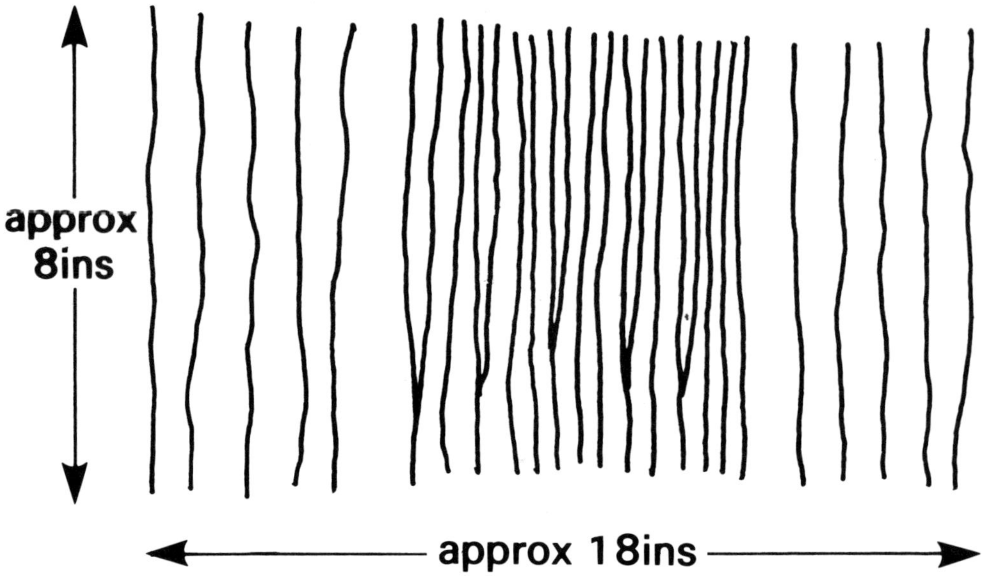

Fig. 17. *Pattern seen on long stone in hedge on Nant-y-Groes Farm.*

II. *Foel Feddau (467m) (Fig. 15)* (145,103324)

1. There is an apparent 070° alignment to Carnalw which was without any doubt merely a directional pointer to some more distant point long lost. (There was in any case a remarkable 070° alignment from neighbouring Foel Cwm Cerwyn which would doubtless have attracted all the attention as the time for the Sun's rising in that direction approached.) The interesting thing is, that it was from Carnalw that the four rhyolite boulders found in Stonehenge were taken (cf. III Carnmenyn). It is to be noted therefore that Carnalw is ⟶ 070 250 ⟵ in relation to Foel Feddau as Carnmenyn is ⟶ 070 250 ⟵ in relation to Freni Fawr. This could well have played a part in the importance of Carnalw (rhyolite) and Carnmenyn (bluestone) from each of which sites boulders were removed to Stonehenge. The importance of their lying on the 070° alignments cannot be over-estimated in assessing their value as potential stones for the building of Stonehenge.

2. The 120°, 130° and 140° alignments have been discussed in the consideration of Foel Cwm Cerwyn as has the 220° and 230°.

3. The 250° February and November sunset alignment belongs to those of very ancient origin. A similar alignment from Foel Cwm Cerwyn cannot be found, might never have existed or could have disappeared. It is not as easy to destroy a hill as to destroy a man-made marker and perhaps it is for that reason that the alignment to Mynydd Castlebythe remains marked and with signs of considerable human activity on its crest and slopes.

4. The same remark holds for the survival of the 270° Equinox alignment to Mynydd Cilciffeth. So, where Man had to compensate for the lack of a suitable 090° Equinox marker not provided by Nature from Foel Cwm Cerwyn by planting his own Standing Stones to this end, nature rose once again to the occasion and provided Mynydd Cilciffeth as a very acceptable 270° Equinox alignment from Foel Feddau.

5. On Carn Ingli Common, the emplacement of Carn Edward provides the direction of the midsummer sunset 310° which is missing from Foel Cwm Cerwyn having either never satisfactorily existed or having been destroyed. This sunset marked the end of the longest day and for this reason was one of considerable importance in the solar calendar.

6. The 320° moonset major northerly standstill alignment to the highest point on Mynydd Carn Ingli and the subject of human marking at that point was probably better and more valued than the similar alignment from Foel Cwm Cerwyn marked lower down on the slopes of Carn Ingli at 311m and perhaps belonging to people of a different generation living at a different time. Judging from the signs of human activity on this 347 metre summit it was in all probability the one and only 320° alignment from Preseli accepted by many generations of people over a great length of time.

III. *Carnmenyn (145325) 364 metres*

Carnmenyn or Carnmeini is one of a group of cairns scattered within a short distance of each other between Foel Feddau and Foel Drygarn. From Carnmenyn there is an 050° reading to Foel Drygarn, an 070° reading to Freni Fawr and a possibly acceptable 270° Equinox alignment to Foel Feddau as well as a 130° alignment to a point on Foel Drych (see Fig. 16). Whether others ever existed cannot be known (see remarks in II (1) of this chapter, page 90).

In addition to being ⟶ 070 250 ⟵ in relation to Freni Fawr, what is distinctive about Carnmenyn is that it is from Carnmenyn that the spotted dolerite or bluestones were taken and hauled all the way to Salisbury Plain. Whether they had more significance over and beyond their undoubted attractive colour, particularly when wet, will never be known. That they were outstandingly venerated for some particular part they played within the astronomic layout of the Preseli observatory, whatever that part may be, whether aesthetic or practical,is certain, and to be on an ⟶ 070 250 ⟵ alignment relationship to Freni Fawr was a famous credential!

The only established fact at the present time is that, of all the possible places in Preseli from which stones could have been removed for use in Stonehenge, it was from Carnmenyn that they were taken. H. H. Thomas (1923) in "*The Source of the Stones of Stonehenge*" says that the limited area from which the stones were obtained and the absence of any apparent reason for the choice of them other than that they must have been held in the greatest veneration and, in addition, the abundance of megalithic remains mainly formed of the same rock-types now to be seen in the Preseli area, are the three important points to remember in considering this subject. In his opinion, the removal of what must have been a venerated stone circle from the eastern end of the Preseli ridge to Salisbury Plain is what took place. According to H. H. Thomas (p. 252) the foreign stones remaining within Stonehenge are thirty-four in number — 29 dolerite masses (bluestones) 4 rhyolite boulders and 1 sandstone. He points out that while dolerite boulders can be found in other areas — such as in Devon and Cornwall and North and South Wales and the South of Ireland, many belonging to the same period of intrusion and exhibiting, in general, certain petrographical characters in common, there are nevertheless, in certain structural features of a macroscopic nature, characteristics that cause the dolerite of Preseli to be distinct specifically from all similar rocks in the British Isles and they are thus easily identifiable even when removed from their point of origin. H. H. Thomas considers that it can be indisputably confirmed that the 29 dolerite masses used in the construction of Stonehenge are all of the same nature and character and that all match in macroscopic and microscopic details with the dolerite masses that outcrop along the Preseli ridge especially in the outcrops of Carnmenyn.

The four rhyolite boulders which are found in Stonehenge are identical in colour, mode of weathering and all structural and mineralogical details with the rhyolite boulders found in Carnalw in the Preseli range — which fact in itself strengthens the conclusive evidence that both types of foreign stone found in Stonehenge (the dolerite bluestones and the rhyolite boulders) originated from the Preseli range. That two types of foreign stone can be proven to have come from Preseli, rather than that only one type, provides in the opinion of H. H. Thomas, incontrovertible evidence that *all* thirty-three of the masses now forming part of Stonehenge came from the eastern end of the Preseli mountains.

It is difficult to reconcile this incontrovertible evidence with the view of G. A. Kellaway (1971) who suggested that ice sheets may have carried the vital Stonehenge foreign boulders to Salisbury Plain.

CHAPTER XV

"A brave new World for ideas as old as the hills"

By about 2,000 B.C. journeying from East to West and West to East was increasing and there had been talk of the existence of a vast plain over in an easterly direction which would lend itself to the setting-up of an observatory. Here, the astronomer-priests on Preseli had learnt, were no topographical features to frustrate, and as long as Man supplied the necessary stones as markers, an observatory could be built that could be the complete answer to their needs in which all the knowledge and experience of millenia, from the time of the simple original and primitive Sun alignments of Drygarn Fawr to the later more intellectually diverting experiments of the Preseli observatory could be brought together to complete fruition in an observatory of unequivocally correct alignments.

R. J. C. Atkinson in "*Stonehenge*" (1956) is of the opinion that the Heel Stone and the ditch at Stonehenge belong to the early phases of the monument's history. What is of interest here is that in 1953, a fragment of freshly fractured bluestone was found in the material used for filling in the ditch and this leads Atkinson to the conclusion that the bluestones were present on the site from early days. While Atkinson puts the Heel Stone in the Stonehenge I phase, he concludes that the ditch cannot be earlier than Stonehenge II (on account of pottery finds associated with it) and this therefore must be the time too, that the bluestones first marked their presence at Stonehenge by the finding of a blue-stone chip in the ditch. This would therefore put the time at approximately 1750-1550 B.C. in the opinion of Atkinson.

While the arrangements for the transfer of the bluestones and rhyolite boulders from Preseli were taking place, it was obvious that the importance of the site on Preseli as an ever developing experimental observatory was coming to an end. The use of the site as a practical observatory however was to continue for some considerable time longer since the need continued for accurate foretelling of the seasons and the worship by now associated therewith.

Plans had, therefore by now, been drawn up for the establishment of another observatory on the less topographically-frustrating site of Stonehenge. With special stones, chosen for the task, venerated for their beauty, especially when wet, and endowed with an aura of mystery as to why this should be, there would be no doubt at all that this observatory would be superior to all previous. And were not all the astronomer-priests who had graduated from an Academy of Learning thousands of years old going to attend on site to their alignment and manner of placing?

This, most surely, was the thinking behind the eventual decision to establish what has in modern terms been categorized as Stonehenge II. Whether Stonehenge ever saw the realisation of these highly ambitious hopes lies outside the scope of this book to consider, but there would seem to be some justification for Burl's comment that in company with the Dodo "both represent a dead-end in development".

Atkinson (1956) in "*Stonehenge*" says that in Stonehenge II there is indeed specific evidence that the Sun played some part and perhaps the central one in the beliefs of the builders. "For the orientation of the entrance of the circle is marked towards the mid-summer sunrise," he continues, "and the monument would, for this reason, seem to represent a new orientation of religious belief."

Evidence from the Welsh hills corroborates the opinion "that the Sun played some part and perhaps the central one in the beliefs of the builders". Hawkins in "*Stonehenge Decoded*" (1970) found alignments for midwinter sunrise, major and minor southerly moon standstills, mid-winter sunset, major and minor northerly moonset and mid-summer sunset. These alignments are by now only too familiar to the reader of this book. He has read of their occurring time and time again in the most primitive and remote regions of the Welsh hills at the same time as those of alignments of importance reaching as far back as Upper Palaeolithic times. The very frequency of their repetition forces the acceptance of their validity on even the most sceptical.

That there should have been "a new orientation of religious belief" might perhaps be better described as the setting up of a new place of worship for a religious belief that had had its origin some thousands of years previous to Stonehenge.

Thus we have seen how Man, as he developed from the uncomfortable life of the sub-arctic tundra into the rather easier but none-the-less strenuous life of the post-glacial period, took with him his anciently acquired knowledge of meaningful Sun alignments and continued his hunting activities uninterruptedly in accordance with them. First, from Preseli (as far as we can tell now, but obviously before Preseli, from some other unrecordable site in Upper Palaeolithic times and perhaps even earlier than that) to Drygarn Fawr, then several millenia later from Drygarn Fawr back to Preseli, then from Preseli to Stonehenge.

That astronomic knowledge and experience was acquired from elsewhere during the course of these thousands of years is highly probable since in the later Mesolithic and early Neolithic periods travel from area to area was becoming ever more common and it is to be expected that discussion and exchange of ideas would have taken place.

By the time 2,000 B.C. had arrived when what is called the Bronze Age was already well established, sites of solar observatories of quite some sophistication were in use in other parts of the country. Wood in "*Sun, Moon and Standing Stones*" (Chapter V) goes so far as to say "there is some evidence that people as far apart as Wales and Caithness were using the same calendar at this period, and that these calendars were actually in synchronism." As far as synchronism is concerned this may or may not have been so, but the fact remains that the peoples of that period were in all probability already quite aware of the existence of sites in other parts of the country where time calendars were kept by observing the movements of the Sun.

CHAPTER XVI

Finale
"Thus, on a level chalk downland in Southern England . . ."

There is an account in "*The Stone Circles of the British Isles*" in which Aubrey Burl describes the first visit he ever made to Stonehenge many years ago during the war. He was filled with wonder as to how it had been possible for people before the mechanically-gifted Romans to have achieved such feats of engineering. To help him, there was a little guide-book with line drawings by one, Heywood Sumner, which told him that the stones had been lifted by *"Megalithic builders from the west* (the italics are mine) who had mingled with vigorous incomers from the east known as Beaker Folk". Burl goes on to say that it did make one very important point in that it said (quoting from Stevens, F. (1938) "*Stonehenge today and yesterday*", London). "Stonehenge may come to be regarded not as an isolated monument, but rather as the climax of a long chain of stone circles . . ." "But it hardly mentioned astronomy," comments Burl, "and certainly had nothing to say about the reasons for designing such an amazing edifice and what it was used for by those remote people on Salisbury Plain."

To the reader of this book, the writer hopes that the word 'remote' would certainly not be appropriate as a description for these "people on Salisbury Plain" with whom he must now, after fifteen chapters, be very well acquainted!

The reader has only to remind himself how, in the opening chapters of this book, he has seen Ancient Man of the Welsh hinterland having to make the best of a new situation offered him by Nature in his effort to survive. In a situation that Nature could not offer him before when the surface of the land was covered by Pleistocene Ice, he has seen how Man, availing himself of what Nature had to offer, her horizons, her hills, her mountain ranges and her Sun, set up a system for survival, the success of which astounds.

With only his eyes to judge, Nature's own horizons to indicate, his own power of memory to remember, short in years and short in stature, he laid the foundation of what astronomy is today.

It has been shown how the setting up of primitive Sun alignments from Drygarn Fawr in the post-glacial era (themselves developing complexities of which the earliest men never dreamed) led to the seeking out of improved fields of more advanced investigation, where the accumulated experience of thousands of years of Sun and Moon observation in the Welsh hill country was transferred to another observatory, that of Preseli. Then, in the realisation of the limitations that even Preseli imposed on progress in astronomy, a move was made yet again. Armed, this time, with even wider experience of a more advanced kind, Man set off once more in the hope and conviction of being able to establish an even more perfect instrument for the examination of the skies on Salisbury Plain.

The reader, being now fully aware of the scale of the undoubted and hitherto unsuspected expertise that these ancient peoples had acquired in the thousands of years of study of sun and moon alignments, long before Stonehenge as such was ever thought of, it is of value to invite for consideration one other final and all important aspect of the subject which cannot fail to exercise the mind.

From Foel Cwm Cerwyn in Preseli to the Rollright Stones in Oxfordshire, 140 miles away as the crow flies, there is a direct West to East line — in other words, *Nature's landmark* Foel Cwm Cerwyn on a reading of 51°57′N 4°46′W is in direct alignment with a *man-made landmark* namely the Rollright Stones at 51°58′N 1°32′W. In the intervening 140 miles, there are several significant *man-made landmarks* on the 51°57′N latitude line as well as several that are topographically occurring. Of the man-made landmarks, let there be mentioned, moving from the Rollright Stones westward, the significantly sited Ring on Cleeve Hill (Glos.) 51°56′N, a stone ring on Mynydd-bach Trecastell (Powys) 51°57′N near the Roman camp Y Pigwn, and nearer to Foel Cwm Cerwyn — Crugbach on 51°57′N and Crugevan also on 51°57′N. There are also, close to this line, Standing Stones 'wayfaring directional' in nature. One of these is of more than ordinary interest and is known as Carreg Fawr 51°57′N approximately. It stands alongside the present A40 some three miles south of Llanwrda and undoubtedly marks the one and only safe fording place of the River Tywi of the relative time as well as being a possible West-East 51°57′N line marker.

If the significance of this 140 mile long West-East alignment, with all its markers both man-made and topographical (these latter also bearing signs of his signature), can be accepted only with reluctance, then Ancient Man is being denied recognition of the fact that he ever executed anything, with deliberacy, and after a deal of deliberation. The only possible conclusion that can be arrived at is that Man had most definitely something in mind whatever it was and that he had had something in mind for quite some considerable time. The length of time that it must have taken to establish such a chain of correctly aligning points is in itself proof enough both of intention and determination.

In order to progress to the production of an effective prototype from the blue-print that he held in his mind, this Man knew that his first task was to find the exact astronomic site, one that he knew to be the one and only in latitudinal and longitudinal terms that would meet the strict requirements that he had in mind.

If the latitude and longitude in which the Rollright Stones were erected actually came into consideration for experimentation is a matter only for conjecture. Described by Burl as "oddly remote" and again as "one of the most famous circles in the British Isles," this circle was called "the greatest Antiquity we have yet seen . . ." by Stukeley, according to Burl. Burl himself suggests that they may have been put up "in the early second millenium" and from his detailed description of this far from simple stone circle, it is fair to deduce that some not inconsiderable astronomic computations or other were at one time here in progress.

Whatever it was that this Man had in mind, it is quite clear that at Rollright all was not right. With the advantage of hindsight and bearing in mind where Man's final choice fell, it is clear that Rollright was slightly too far East and even more too far North.

Hawkins and others have already suggested that the site of Stonehenge was actually chosen and that would, of course, have had to have been after long experimentation — not only at Stonehenge itself but at sites in the close latitudinal and longitudinal vicinity as well. It is of interest to study Fig. 1 on page 9 in Burl's "*The Stone Circles of the British Isles*" in which the West to East line from Preseli to the Rollright Stones can be seen as is also the progression southward via Devil's Quoits then westward again slightly toward Winterbourne Basset 51°29′N 1°50′W then southward 'a touch' to Avebury 51°26′N 1°52′W and thence to Stonehenge 51°11′N 1°50′W. It needs to be pointed out however that in Fig. 1 in Burl's book is intended merely as a record of "The Distribution of Stone Circles in the British Isles" and relates in no way to possible experimentations in the finding of a site answering certain precise requirements.

It was Newham who in 1963 drew attention to the fact that the latitude of Stonehenge is virtually unique. What is unique about it is that the extreme northern and southern risings and settings of the sun and moon are at right angles to one another, whereas, a few miles north or south, such cross alignments would be along the sides of what he called "a skewed parallelogram." Nor was this all. It has been pointed out moreover that the diagonal, across the rectangle so formed, indicates "those sunrises and sunsets falling between the solstices and the equinoxes" — those points in time of early February, early May, early August and early November — times of the year of importance to Upper Palaeolithic Man, inherited and revered by his Mesolithic descendants and continued and passed on to be recorded in the primitive sun calendars of the Welsh Uplands and the more sophisticated solar observatories of the Preseli complex. The are seen as still of great importance in the long experimentation by Mesolithic Man's descendant, Neolithic Man in his patient, arduous and dedicated search for the site of all sites, the site from which could be brought to reality that dream, that mental blue-print, in which the extreme northerly and southerly risings and settings of the sun and moon would be at right angles to one another thus providing most importantly a situation from which it would arise that the diagonal across the rectangle so formed would honour, enshrine and provide those points in time between the solstices and the equinoxes that the peoples of the previous eight thousand years, if not longer, had seen fit to leave recorded for ever on those wild hill-tops of the remote fastnesses of Wales.

Little wonder then that of all the stone circles of the British Isles, Stonehenge was considered peerless and worthy of special attention. As far as is known at the present time no other circle in the British Isles contains stones that were brought more than a few miles at the most, and to no other circles in England do bluestones from Preseli seem to have been taken.

Thus, on a level chalk down-land in Southern England, did Man produce artificially with massive stones and columns, the important celestial alignments which, occurring as natural topograpical features in upland Wales, he had used for so many long years previously.

The mighty mountains stay,
  The mighty winds unfold;
With each new dawn is heard
  The shepherd's song of old.
At the foot of crag and hill
  The little daisies grow,
New shepherds now are seen
  On these mountains as of yore.

English translation of Ceiriog by Rhydwen Williams

Recommended Reading

Archeologica Cambrensis

Arch. Cam. II Bushell, W.D. (1911) p. 287-333. "Among the Prescelly circles".

Arch. Cam. Thomas, Roger (1912) p. 211-216. Flint factory in Aberystwyth.

Arch. Cam. (1923) p. 155. The Abergwesyn flint factory.

Atkinson, R.J.C.
"Stonehenge" (1956) Hamish Hamilton.
"Some new measurements on Stonehenge" Nature 275 (1978).

Barber, Chris.
"Mysterious Wales" (1982) David & Charles.

Bowen, D.Q.
"The Llanelli Landscape" (1980) Llanelli Borough Council

Burgess, Colin
"The Age of Stonehenge" (1980) J. M. Dent & Sons (402pp).

Burl, Aubrey
"The Stone Circles of the British Isles" (1976) Yale University Press. 410pp.

Clark, J.G.D.
"The Mesolithic Settlement of Northern Europe" (1936) Cambridge University Press. 284pp.

Duhme, Hans
"Döhlen dat ole Dorp" (1982) Heinz Holzberg Verlag. 223pp.

Evans, J.G.
"The Environment of Early Man in the British Isles" (1975) Paul Elek, London.

Gooch, Stan
"Guardians of the Ancient Wisdom" (1979) Fontana Collins. 287pp.

Hawkins, G.S.
"Stonehenge Decoded" (1970), Fontana, London.
Hoyle, Fred
"Stonehenge — an Eclipse predictor" (1966) Nature 211. pps. 454-456.
"Speculations on Stonehenge" (1966) Antiquity 40. pp.272-276.
John, Brian S.
"Pembrokeshire" (1984) Greencroft Books, Newport (Pembs) 208pp.
"Presely Hills" (1981).
Kellaway, G.A.
"Glaciation and the Stones of Stonehenge" (1971). Nature 232 No. 5314. 30-35.
Laing, Lloyd and Jennifer
"The Origins of Britain" (1982) Granada Publishing Ltd. 256pp.
Lewis, E.T.
"North of the Hills" (1972) (Haverfordwest).
"Llanfyrnach Parish Lore" (Haverfordwest).
Lockyer, N.
"Stonehenge and other British monuments astronomically considered" (1906) London, 95pp.
Murray, Margaret
"The Witch Cult in Western Europe" (1962) Oxford University Press.
Newham, C.A.
"The Astronomical Significance of Stonehenge" (1972) Blackburn, Leeds.
Stonehenge — A Neolithic "Observatory" (1966) Nature Vol. 211, 456-457.
Thom, A. and Thom A.S.
"Megalithic remains in Britain and Brittany" Clarendon Press, Oxford.
Thomas, H.H.
"The Antiquaries Journal" Vol. 3 (1923).
"The Sources of the Stones of Stonehenge" 239-264pp.
Vergilius, Maro P.
"The Aeneid" Book III (19 B.C.) Line 19-21.
Published posthumously under the tutelage of the Emperor Augustus.
Watkins, Alfred
"The Old Straight Track" (1974) Abacus. 234pp.
Wood, John Edwin
"Sun, Moon and Standing Stones" (1980) Oxford University Press. 217pp.

GEDOE? VALLEY LIBRARIES

C		MT	
AC		P	~~8. SEP. 1986~~
CB		A	
H		Y	- 8. SEP. 1986
M			